The Mathematics of Great Am

The Mathematics of Great Amateurs

SECOND EDITION

JULIAN LOWELL COOLIDGE

with an introductory essay
by

JEREMY GRAY
The Open University

CLARENDON PRESS · OXFORD
1990

Oxford University Press, Walton Street, Oxford OX2 6DP

Oxford New York Toronto
Delhi Bombay Calcutta Madras Karachi
Petaling Jaya Singapore Hong Kong Tokyo
Nairobi Dar es Salaam Cape Town
Melbourne Auckland

and associated companies in
Berlin Ibadan

Oxford is a trade mark of Oxford University Press

Published in the United States
by Oxford University Press, New York

British Library Cataloguing in Publication Data
Coolidge, Julian Lowell, 1873–1954
The mathematics of great amateurs.
1. Mathematics—Biographies—Collections
I. Title
510'.92'2
ISBN 0-19-853939-8

Library of Congress Cataloging in Publication Data
Coolidge, Julian Lowell, 1873—1954.
The mathematics of great amateurs / Julian Lowell Coolidge: with
an introductory essay by Jeremy Gray.—2nd ed.
Includes bibliographical references.
1. Mathematicians. 2. Mathematics—History. I. Title.
QA28.C6 1990   510'.9—dc20   89—23110
ISBN 0-19-853939-8

Set by Joshua Associates Ltd, Oxford
Printed in Great Britain by
Biddles Ltd, Guildford & King's Lynn

FOREWORD

Coolidge's book is something of a classic in the history of mathematics, although it has not inspired others to increase the list of mathematical amateurs as Coolidge had hoped it would. It owes its status to its remarkable breadth of erudition, and its inimitableness to Coolidge's style of writing. This is cheerfully personal, some might say obtrusive, and not at all the manner of the usual self-effacing historian. Coolidge frankly confesses when he does not understand something, and is not above suggesting that some ancient writing is simply unintelligible. In the end, this openness becomes part of the book's charm. For that reason amongst others this re-edition of the book is a textually unaltered reprint of the 1949 Oxford University Press edition, which coincides with the 1963 Dover reprint.

None the less, and I believe in keeping with Coolidge's own love of scholarship, I have sought to bring the book up to date by writing an introductory essay. The first part of this essay is a short biographical account of Coolidge himself, based on Struik's obituary of him. The second part of the essay deals in turn with each of the 16 chapters of the book. I have tried to describe substantial pieces of information not available when Coolidge wrote, and I have indicated where later scholars have wanted to disagree significantly with Coolidge, either on the basis of new information or a radically different interpretation of the evidence. Minor cavils will be found in the few new footnotes I have added at the end of the book. There is also an additional, short, and selective bibliography which readers can use if they wish to consult modern scholarship on some questions.

1989   J.J.G.

PREFACE TO THE FIRST EDITION

THE responsibility for advancing our knowledge in all branches of science has always in the past, and still more in recent times, rested mostly on the shoulders of men professionally engaged in the various subjects. The reasons for this are partly social, but mostly financial. The extreme cost of the apparatus needful for study in the physical and other experimental sciences, and the large collections needed in the biological sciences, presuppose an expenditure of money which places the conduct of research in such subjects beyond the means of men not professionally connected with institutions which have large resources available. Even in mathematics, where there is little need for equipment, advanced study is open only to those who have easy access to extensive library facilities, and these imply amassed wealth. Nevertheless, mathematics is better off than the other sciences in this respect, and throughout the centuries there have been a certain number of men, not professional mathematicians, who have made significant contributions to this, the oldest of the sciences.

It has seemed to me worth while to make some study of the contributions of these men whom, for want of a better term, I have called amateurs. But I discovered at the outset that a rigid definition of this term was not feasible and a consistent policy as to who should be included and who should not, practically impossible. One would naturally mean by an amateur in mathematics one who did not earn his living in large part by the subject, as a teacher or a physicist, or an astronomer or even an engineer. But under such a definition Euclid and Archimedes would be classed as amateurs, which seemed to me absurd. On the other hand, I did not like to leave the Greeks out altogether, so I have included Plato.

In general I have taken men who were principally known for some other activity, yet whose success in the field of mathematics enabled them to make contributions of permanent value. As I said before, it has been beyond my strength to be consistent. I have included Omar Khayyám although he was an astronomer, but he was so widely known and loved as a poet, thanks partly to Edward Fitzgerald, that I could not bear to leave him out. I have regretfully excluded Sir Christopher Wren, that prince of architects, and Charles Lutwidge Dodgson, the writer of the most popular humorous work in generations, because unfortunately the former was Savilian Professor of Astronomy in the University of Oxford, and the latter was mathematical tutor in Christ Church in the same institution. I have included Baron Napier whose non-mathematical activity amounted to little beyond acrimonious theological writing because the discovery of logarithms was of such transcendent importance. I have not included Fermat, whom Bell has called the Prince of Amateurs,† who was a 'Maître des requêtes', because he was so really great that he should count as a professional, but I have included the great philosopher Pascal. I have included William George Horner, no great mathematician, because it seemed to me interesting that a man of very limited education, beyond his own reading, should discover, independently of previous discoverers, what is still the best method of solving an age-old problem. I have not included George Salmon, a theologian, who is said to have forgotten his mathematics in later life, because his text-books were so extremely good as to place him in the professional ranks, nor Sir Thomas Heath, joint permanent Secretary of the Treasury, or Paul Tannery, 'Ingénieur des Tabacs', because the work of these fine scholars was in the history of mathematics where I did not feel myself competent to appraise. If consistency is a vice of small minds, it is a vice I have successfully avoided. My friend Professor Archibald of Brown University, who has been very helpful to me in the preparation of this work, truly remarked that the number of men included could easily be doubled or trebled. I should be most happy to see someone undertake this interesting task.

CAMBRIDGE, MASS.   J. L. C.

† *Men of Mathematics*, ch. iv (New York, 1937).

CONTENTS

BIOGRAPHICAL NOTE

Julian Lowell Coolidge was born in Brookline, near Boston, on September 28, 1873. His family descended from John Coolidge, one of the first settlers of Massachusetts Bay, as did that of President Calvin Coolidge in Vermont. They were people of distinction in the area, and gave their son a good education that culminated in his studying mathematics at Harvard and then at Oxford. He returned to teach mathematics at the school in Groton, which was then directed by its founder the Reverend Endicott Peabody. One of his pupils was Franklin D. Roosevelt, with whom he remained on friendly terms. In 1899 Coolidge became an instructor in mathematics at Harvard, and in 1902 a member of Faculty. That year he left for two years study abroad, where he took a Ph.D. from Bonn under the direction of Eduard Study. He took the opportunity to travel to Turin and work under Corrado Segre, and these two leading geometers were a decisive influence on him. Study had just written an important book on line geometry in non-Euclidean space, a topic that derived from the theory of rigid body mechanics and projective geometry in equal measure and that had been advanced by such men as Plücker, Klein, and Lindemann. Segre was the leading Italian geometer of his day, an expert in the birational geometry of curves and a formidably well-read scholar. Many of Coolidge's strictly mathematical books and papers bear the mark of his time with Study and Segre.

In 1918 Coolidge became a full professor at Harvard, and in 1929 the first master of Lowell House, President Lowell having just introduced the house system to Harvard. His tenure there was a success, and he stayed as professor and master until he retired in 1940. He was a good teacher, and a great lover of the energetic life; he once held the record for the mile at Harvard (4 mins, 30.8 seconds). As an older man he transferred his energy to fund raising for the American Mathematical Society and the Mathematical Association of America. He advocated the conservation of natural resources and forested a large tract of woodland near his home in Maine.

Judged by the highest standards, he was not among the leading mathematicians of his day, and he often seems to have thought that his own field of study had entered into an irreversible decline. This was far from true, and if his own contributions now seem dated they were valuable in their time, not least for the education of mathematicians in America. Starting in his sixties he wrote three historical books, of which this, the last, was written in retirement when he was 76. They are all personal and lively statements, mostly devoted to geometry, the topic that had caught and held his interest all his life; they too deserve to remain in print. He died on 5 March 1954.

J.J.G.

INTRODUCTION

Jeremy Gray

There is much in Coolidge's book that one can still admire, and a great deal not readily accessible anywhere else. This is surely the result of his enthusiasm for reading the original texts coupled with a respect for the best historical writings he could find. But if, like a well-built house, his book has stood the test of time, it none the less now stands in need of some repair.

Our problems with it start with the title, and Coolidge's avowedly idiosyncratic use of the word 'amateur'. In recent years historians have paid growing attention to the working situation of mathematicians and scientists. Terms like 'amateur', 'professional', and 'patron' have come to be used precisely and effectively to elicit aspects of scientific work: why it was done in this place or that, why it was done in this way or that. One should not be misled by the title into placing Coolidge's work in a tradition that has only grown up since its author's death. Work on 'high' and 'low' traditions in mathematics, with the concomitant choice of Latin or the vernacular in some periods, and on the diffusion of numeracy, likewise postdates this book. But the reader should have no difficulty placing what he reads here into those more general contexts, and may find on occasion that it helps bring them more carefully into focus.

Coolidge also renders the ancient mathematics in a way that has come in for considerable criticism. Many would now argue against the tradition that found no problem turning the prolixity of older writers into elementary algebra. Those who, like Coolidge, did so believed that they were reading between the lines and faithfully presenting what was clear to the original author in a way that is clear to the modern reader. Modern historians, in contrast, would emphasize the distinctive ways of thinking of the older writers, and argue that it is the way they differ from us that made them find certain arguments natural and others, perhaps, impossible. The least that needs to be said is that no-one before Descartes literally wrote as Coolidge might seem to imply they did.

The essay on Plato shares the high view then current of the achievement of the Pythagoreans. Recently, scholars have questioned some of the views of Heiberg and Heath, which rest on surprisingly little contemporary evidence. Indeed, one of the ways in which Plato's writings are important is that they are amongst the earliest reliable evidence we have for the development of Greek mathematics. Much else must be quarried carefully out of Euclid's *Elements*, a later and harder source. The discovery of incommensurability has been much discussed by historians; a spread of contemporary views will be found in the books by Burkert, Fowler, Knorr, and Szabo listed in the additions to the bibliography. Coolidge's tacit disagreement with those who spoke of a foundational crisis in Greek mathematics consequent upon this discovery is in keeping with the later work. A further interesting source, not mentioned by Coolidge, is that of the relation between mathematics and music; this has been well described by Barker [1984]. On the history of the regular solids, mention should be made of Waterhouse's important observation that, mathematically, the important discovery was not that of the fifth solid (the dodecahedron) but the concept of regular solids and the idea that they could be completely classified.

Scholarship has also found fault with some of Coolidge's translations from the Greek, and alternatives to most of them can be found in the invaluable two-volume reader edited by Bulmer-Thomas [1939]. Only one of these, however, seems to be egregiously awry, the one that straddles pp. 12-13. A better version is provided in a footnote.

The essay on Omar Khayyám does not begin with a biography modern authorities would like; those provided by Berggren, Jaouiche, and Rosenfeld are fuller and more accurate. Berggren in particular argues forcefully that the story of the childhood pact with Nizam al-Mulk cannot be trusted. Coolidge's opinion of Fitzgerald as a translator is likewise contested; good poetry though it is, it is not very faithful to Khayyám's original and the version of Arberry is often markedly different. That said, the account Coolidge gives of Khayyám's work on cubic equations is typical of his virtues as an expositor (note being taken of his policy on the use of modern symbolism).

The topic which Coolidge would surely wish to discuss if he were writing the book today is Khayyám's remarkable study of the parallel postulate in Euclid's geometry. This was put into English (Amir-Moez, [1959]) and has since been much written about. It is a trenchant disagreement with Khayyám's illustrious predecessor, ibn al-Haytham, over the use of the concept of motion in proving theorems in geometry. Ibn al-Haytham had sought to defend the parallel postulate by arguing that if a line segment moves so as always to be perpendicular to a given line, then its tip sweeps out a straight line parallel to the given line. This granted, he could establish the truth of the postulate. Khayyám did not dispute the postulate, but he rejected ibn al-Haytham's method and offered one of his own based on a dictum about straight lines that he attributed to Aristotle (and which has not otherwise come down to us). Khayyám's method also fails, as it must, to establish the truth of the postulate, but this disagreement about the status of motion in geometry strikes deep into the study of what geometry is about. For a full discussion of the work of ibn al-Haytham and Omar Khayyám, and of the centuries-long Islamic investigations of the foundations of geometry, the reader may consult the works by Jaouiche (where he will also find French translation of many originals) and Rosenfeld, or even my own book, which is at this point based on theirs.

The three chapters on the work of the artists Piero della Francesca, Leonardo da Vinci, and Dürer stand most in need of rethinking from a social historical standpoint. When they discuss perspective, they should be set in the context established by White's excellent book *The birth and rebirth of pictorial space* [1957]. That on Piero should be read in conjunction with M. D. Davis *The mathematics of Piero della Francesca* and Jayawardene [1976], where another of Coolidge's omissions is also filled: that of Piero as an algebraist. It emerges that Piero's treatise was based on an earlier one by Fibonacci and so dwelt heavily on the solution to quadratic equations, but went on to discuss a few special equations of higher degree. It is also amusing to note that when discussing whether or not Pacioli stole the material for his book *De divine proportione* from Piero, Coolidge does not observe what is now widely agreed, that the beautiful drawings in that book are the work of Leonardo.

It is hard not to write about Leonardo without resorting to superlatives, and Coolidge is no exception. One wayin which doing so keeps us from appreciating this remarkable man is demonstrated by contrasting the account here with the ones by Gombrich and Kemp which introduced the catalogue of the 1989 exhibition *Leonardo then and now*. Gombrich points out that to praise Leonardo for uniting art and science would be to use the terms in a way that Leonardo would scarcely have understood. Leonardo, he argues, emphasized that painting had to rest on knowledge and so be valuable as a liberal art and not merely as a craft. This was not a snobbish aspiration for status, but part of Leonardo's belief in the importance of artistic creativity. Understanding the laws of nature is scarcely part of the artist's rationale today. Nor, Gombrich argues, should Leonardo's universality be over-emphasized. Leonardo considered himself unlettered because he never really mastered Latin and had been brought up not in the schools of Renaissance humanists but instead in the popular, vernacular tradition. His marvellous powers of observation can also be allowed to hide him from our sight. Truesdell amongst others has drawn attention to the fact that Leonardo belongs with those who looked and experienced, not those who experimented. Many of his drawings of vortices, for example, amply convey the vigorous swirls such things have, but cannot be said to be literally accurate. Leonardo's style of representation is neither that of photographic realism nor is it adequately experimental, still less theoretical.

The chapters on Leonardo and Dürer none the less stand up favourably with the treatments each received in the *Dictionary of scientific biography*. By relying on Marcolongo, Coolidge put his trust in someone whose work on Leonardo has not been surpassed. For a rich account that connects Leonardo's study of mathematics with the rest of his work, see M. Kemp [1981].

The material on Napier likewise stands up well. The modern historian, who expects to see geometry and not algebra in works of the period, does not find Napier's presentation of the idea of logarithms as alienating as does Coolidge, who otherwise brings out Napier's ideas very clearly. He might have added that Napier's work excited the admiration of no less an astronomer than Kepler.

The description of the invention of decimal fractions has, however, been overtaken. Reasonably enough in such a small, tightly focused book, Coolidge did not mention the Chinese method of writing decimal fractions, nor that of the Islamic astronomer al-Kashi (d. 1436). In Europe, Regiomontanus' table of tangents, published in 1561, presented values as parts of 10^5 in what was to become Napier's notation (one lacking only the decimal point). Viète's *Canon mathematicus* of 1579 likewise had the idea of decimal fractions; the notation again lacks only the decimal point. The idea of decimals was first explained clearly by the influential Dutch mathematician Simon Stevin in his book *De Thiende* of 1585, which was translated into English in 1608. Stevin's own notation was strangely cumbersome, and was simplified by Magini when he presented his book of trigonometric tables in 1592. Here the decimal point appears as a comma. Clavius occasionally did the same in his book of tables in 1593. Finally Jost Bürgi, whom Kepler considered to be the inventor of logarithms, used decimals in unpublished work some time after 1592. Since Napier, as Coolidge notes, 'received much of his education on the continent of Europe', it is likely that some of these ideas were known to him. On the other hand later writers have agreed with Coolidge that Bonfils' system was indeed soon forgotten.

We know a little more about Pascal since Coolidge wrote about him. Taton has illuminated our account of his work as a geometer, drawing on surviving notes by Leibniz of the still-lost *Traité des coniques*; see also Field and Gray [1987]. A. W. F. Edwards has written a thorough historical study, *Pascal's arithmetical triangle* [1987], looking not only at the precursors of the Triangle but at Pascal's work on probability. He agrees with Freudenthal [1953] that Coolidge should not have been worried by Vacca's letter into arguing that Pascal may have taken the idea of complete induction from Maurolico; it is not there. Edwards agrees with Rabinovitch [1970] that something more like it is, however, to be found in the writings of Levi ben Gerson, where it is impeded by a poor notation. So it seems that Pascal was indeed the first to provide clear and explicit examples of mathematical induction.

On the other hand, I regret that I have been able to find out nothing about Arnauld. The *Dictionary of scientific biography* article is much less informative on him than was Coolidge.

Jan de Witt and Hudde have been written about by Dutch scholars. Struik provides a good general account of this, the most dramatic period in the history of science in his native Holland, in his *The land of Stevin and Huygens*. H. H. Rowen's book [1978] is a full-length political biography of de Witt, with an interesting chapter ('The unphilosophical Cartesian') on his dealings with van Schooten. J. van Maanen's book [1987] sheds considerable light on the rise and fall of Dutch mathematics in the seventeenth century. That there were both Dutch mathematicians and an audience for their work in places like Leiden can be attributed to the remarkable success of the Dutch in driving out the Spanish in a heroic and often costly war fought from 1568 to 1648. As happened with the French Revolution, this struggle greatly invigorated Dutch intellectual life and brought the Netherlands energetically to international prominence. The reputation the country acquired for religious tolerance attracted Descartes, who lived in the country from 1628 to 1649, and published his *Geometrie* in Leiden in 1637. This difficult book was republished with additional commentaries in a Latin translation by Frans van Schooten in 1649, and that edition was further extended by van Schooten in 1659-1661. It is here, for example, that Hudde's essay appears.

The short chapter on Hudde fails to point out that the reason he presented a method for finding out when a polynomial equation has equal roots is that this is the central problem in carrying out Descartes' method for finding a tangent to a given curve at a given point. Descartes had argued in his *Geometrie* that to find the normal (and hence the tangent, which is at right angles to it) circles are drawn through the given point, having their centres for convenience on the *x*-axis. When the centre of the circle lies on the normal to the curve, the circle will touch the curve; nearby circles will meet the curve in the given point and another nearby. So to find the normal, let the circle centre $(s, 0)$ be drawn through the given point (a, b), and eliminate either x or y from the equation of this circle and the curve. The resulting equation, shall we say in x, has repeated root $x = a$ precisely when the point $(s, 0)$ lies on the normal. This yields an equation for s and so finds the normal and hence the tangent. Unfortunately, the elimination is not easy, and can lead to a complicated equation. Hudde's rule gave a useful way of proceeding. When the young Newton came across it there he speedily saw how to extend the argument to find the centre of curvature of curves. By applying his simplification of Hudde's rule, Newton was also led to ask and answer the 'right' question of a tangent to curves: not 'where does it meet the *x*-axis' but 'what is its slope?'. So Hudde's rule was a catalyst in the discovery of the differential calculus.

There is little to add to the chapter on Brouncker. Coolidge might have mentioned his involvement in the British response to Fermat's challenge of 1657: find integer solutions to the misnamed Pell equation, $x^2 = Ay^2 + 1$ for various specified integers A; among those Fermat proposed is $A = 109$, for which the smallest value of y is the astonishingly large number 15 140 424 455 100. Plainly he had a general method for finding solutions. Wallis replied with a solution method that works for all square-free values of A, and attributed it to Brouncker. It seems impossible to determine which man discovered what, but it is certain that neither they nor Fermat knew of the solution proposed by the twelfth century Indian mathematician Bhaskara. As for the story of the cycloid, D. T. Whiteside [1969, 390-9] finds that Huygens made his discoveries in December 1659 and no doubt mentioned it when he visited London in 1661. At all events, it was a topic of discussion in London in 1662, when Brouncker published his note. Apparently Huygens was not impressed with it, and remarked that it was easy to discover what Brouncker had found once Wren's results on the cycloid were available.

The Marquis de L'Hospital first met Johann Bernoulli, six years his junior, in late autumn 1691, when he was 30. They met at one of the weekly gatherings organized in Paris by Father Malebranche, and Bernoulli quickly impressed L'Hospital by his ability to find the radii of curvature of arbitrary curves. This inspired him to hire the young Swiss as his private tutor for four lessons a week. The lessons continued to the middle of June 1692, when the L'Hospitals left Paris and took Bernoulli with them to Oucques. The terms of the contract stipulated not only the rates of pay, but above all that Bernoulli communicate his discoveries to his new master while concealing them from everyone else. To modern eyes, these circumstances incline one to agree with Bernoulli that he was shabbily acknowledged by the Marquis, but doubtless the nobleman was acting as he would towards anyone whose services he hired. For more details see Spiess [1955], partially translated into English in Fauvel and Gray [1987]. Bernoulli's discovery of the so-called L'Hospital's rule is documented in his letter of 22 July 1694; see Spiess, no. 28, p. 235.

In Book IX of his *Traité* L'Hospital deals with the use of conic sections to solve equations of higher degree, a topic, as Coolidge rightly remarked, that was dear to Descartes. In two important recent papers ([1981], [1984]), H. J. M. Bos has shown just how much of the *Geometrie* must be seen as an attempt to provide geometric answers to geometric questions, a process in which algebra plays the role of a medium. Geometric problems may be interpreted algebraically, but their solutions must then by re-interpreted geometrically. What L'Hospital, like many of his contemporaries, was doing in his chapter on the 'construction of equations' was to solve equations geometrically not because other solutions were unavailable but because geometrical solutions were required. As Bos has shown, this programme ultimately foundered because it could not be given agreed standards for simplicity, and by Euler's time it had become a curiosity. However, Coolidge's algebraic style of writing should not obscure the importance to his contemporaries of what L'Hospital was trying to do.

There is a thorough, recent treatment of the Saint Petersburg Paradox in Jorland [1987]. Jorland not only gives Buffon's table of actual results on playing the game 2048 times, but provides a detailed account of the discussions of many others on this intriguing paradox: N. Bernoulli, d'Alembert, Condorcet, Laplace, and others. The reader will enjoy both this essay and that of M. Paty [1988] which is more informative on Diderot's work on probability. Coolidge did not see those papers by Diderot, and therefore missed his polemics against d'Alembert, which included his attempt to rebut d'Alembert's arguments in favour of innoculation. Coolidge also seems to profess ignorance of Brook Taylor. For an account of his work on the vibrating string and its connection to his interest in music, the reader can do no better than to turn to Cannon and

Dostrovsky [1981]; these authors also discuss Diderot's use of the work of Sauveur on music and mathematics. On the other hand, Coolidge's account of Diderot's other work remains unsurpassed.

I have nothing to add to the chapter on Horner. The chapter on Bolzano is being reprinted at a time when he is again receiving the attention he deserves. Oxford University Press are bringing out a two-volume edition of his mathematical work, edited by S. B. Russ, and I am grateful for his comments. He tells me that the two important principles mentioned by Coolidge on p. 195, the definitions of continuity and of convergence of a series, can be found in an earlier work by Bolzano, the *Binomische Lehrsatz* of 1816. The *Functionenlehre*, mistakenly called by Coolidge the *Funktionentheorie*, is being re-edited from manuscript by van Rootselaar for the *Gesamtausgabe* edition, and the English reader must await its appearance in Russ's book. Surprisingly, Coolidge does not mention Bolzano's splendid example of a function continuous at every point of an interval but differentiable nowhere, which predates the examples due to Riemann and Weierstrass by 20 to 30 years.

Bolzano's function is defined by an iterative process on the interval $[0, 1]$. The function f_0 sends x to x. The function f_1 is defined differently on each of the four intervals $[0, 3/8]$, $[3/8, 1/2]$, $[1/2, 7/8]$, and $[7/8, 1]$; its graph is obtained by joining up the points $(0, 0)$, $(3/8, 5/8)$, $(1/2, 1/2)$, $(7/8, 9/8)$, and $(1, 1)$ by straight line segments. The function f_2 is obtained by replacing each of these segments by four more. The one from (a, a') to (b, b') is replaced by the zig-zag line joining these points: (a, a'), $(a + 3/8(b-a), a' + 5/8(b'-a'))$, $(1/2(a+b), 1/2(a'+b'))$, $(a + 7/8(b-a), a' + 9/8(b'-a'))$, (b,b'). The function f_{n+1} is obtained from the function f_n in the same way. Bolzano's function f is the point-wise limit of the $f_n's$.

Bolzano gave an imperfect proof that the function is everywhere continuous, because he missed the uniformity of the convergence of the sequence f_n. But he did prove that the function failed to be differentiable at a dense set of points and that it is never monotonic on any interval. Modern commentators like Rychlik have filled in the gaps; this treatment is taken from Jarnik [1981], who gives a careful proof of all these claims.

Coolidge would surely also have enjoyed Bolzano's attempts to define the concept of line or curve, surface, solid and continuum, which make him an important precursor of the modern theory of dimension. These have been well discussed by Johnson [1977].

I would like to acknowledge the help of Judith Field, David Fowler, Jan van Maanen, and Steve Russ in preparing these notes. I am, of course, responsible for any errors they may contain.

BIBLIOGRAPHY

Arberry, A. J. (1952). *The Rubaiyat of Omat Khayyam: a new version based upon recent discoveries.* John Murray, London.

Barker, A. (1984). *Greek musical writings, I: the musician and his art.* Cambridge University Press, Cambridge.

Berggren, J. L. (1986). *Episodes in the mathematics of medieval Islam.* Springer Verlag, New York.

Bos, H. J. M. (1981). On the representation of curves in Descartes' *Geometrie, Archive for History of Exact Sciences,* 24, 295-338.

Bos, H. J. M. (1984). Arguments on motivation in the rise and decline of a mathematical theory; the 'Construction of equations', 1637-ca.1750. *Archive for History of Exact Sciences,* 30, 331-80.

Bulmer-Thomas (1939). *Greek mathematical works,* 2 volumes. Harvard University Press and Heinemann, Cambridge, Mass., and London.

Burkert, W. (1972). *Lore and science in ancient Pythagoreanism* (tr. E. L. Minar, jr.). Harvard University Press, Cambridge, Mass.

Cannon, J. T. and Dostrovsky, S. (1981). *The evolution of dynamics: vibration theory from 1687 to 1742.* Springer Verlag, New York.

Davis, M. D. (1977). *Piero della Francesca's mathematical treatises.* Longo Editore, Ravenna.

Edwards, A. W. F. (1987). *Pascal's arithmetical triangle.* Charles Griffin, London and Oxford University Press, New York.

Fauvel, J. G. and Gray, J. J. (1987). *The history of mathematics—a reader.* Macmillan, Basingstoke.

Field, J. V. and Gray, J. J. (1987). *The geometrical work of Girard Desargues.* Springer Verlag, New York.

Fowler, D. H. (1987). *The mathematics of Plato's Academy: a new reconstruction.* Clarendon Press, Oxford.

Freudenthal, H. (1953). Zur Geschichte der vollständigen Induktion. *Archives Internationales d'Histoire des Sciences,* 17-57.

Gombrich, E. H. (1989). Preface to the catalogue of the Leonardo exhibition. Hayward Gallery, London.

Gray, J. J. (1989). *Ideas of space: Euclidean, non-Euclidean, and relativistic,* 2nd edition. Oxford University Press, Oxford.

Jaouiche, K. (1986). *La théorie des parallèlles en pays d'Islam.* Vrin, Paris.

Jarnik, V. (1981). *Bolzano and the foundations of mathematical analysis.* Society of Czechoslovak Mathematicians and Physicists, Prague.

Jayawardene, S. S. (1976). The *Trattato d'Abaco* of Piero della Francesca. In *Cultural aspects of the Italian Renaissance: essays in honour of Paul Oskar Kristeller.* C. Clough, Manchester.

Johnson, D. M. (1977). Prelude to dimension theory: the geometrical investigations of Bernard Bolzano. *Archive for History of Exact Sciences,* 17, 261-95.

Jorland, G. (1987). The Saint Petersburg paradox. In *The probabilistic revolution, I, Ideas in history,* (L. Krüger, L. J. Daston, M. Heidelberger (eds.)), pp. 157-190. MIT Press, Cambridge, Mass.

Kemp, M. (1981). *Leonardo da Vinci: the marvellous works of nature and man*. Dent, London.

Kemp, M. (1989). Leonardo then and now. In the catalogue of the Leonardo exhibition, Hayward Gallery, London.

Khayyám Omar (1959). Discussion of difficulties in Euclid, tr. A. R. Amir-Moez. *Scripta mathematica*, 24, 275-303.

Knorr, W. R. (1975). *The evolution of the euclidean* Elements. Reidel, Dordrecht.

Maanen, J. A. van (1987). *Facets of seventeenth century mathematics in the Netherlands*. Drukkerij Elinkwijk BV, Utrecht.

Paty, M. (1988). D'Alembert et les probabilités. In *Sciences à l'époque de la révolution Française: recherches historiques* (R. Rashed (ed.)). Blanchard, Paris.

Proclus (1970). *A commentary on the first book of Euclid's* Elements, tr. G. R. Morrow. Princeton University Press, Princeton.

Rabinovitch, N. L. (1970). Rabbi Levi ben Gerson and the origins of mathematical induction. *Archive for History of Exact Sciences*, 6, 237-48.

Rosenfeld, B. A. (1988). *A History of non-Euclidean geometry*. Springer Verlag, New York.

Rowen, H. H. (1978). *John de Witt, Grand Pensionary of Holland, 1625-1672*. Princeton University Press, Princeton.

Spiess, O. (1955). *Der Briefwechsel von Johann Bernioulli*, I. Birkhäuser, Basel.

Struik, D. J. (1955). Julian Lowell Coolidge, in memoriam. *American Mathematical Monthly*, 62, 669-82.

Struik, D. J. (1981). *The land of Stevin and Huygens*, 3rd edition. Reidel, Dordrecht.

Szabo, A. (1978). *The beginnings of Greek mathematics*. Reidel, Dordrecht.

Taton, R. (1964). *L'Oeuvre scientifique de Pascal*, various editors. Presses Universitaires de France, Paris.

Waterhouse, W. C. (1972). The discovery of the regular solids. *Archive for History of Exact Sciences*, 9, 212-21.

White, J. (1967). *The birth and rebirth of pictorial space* (1st edition 1957). Faber and Faber, London.

Whiteside, D. T. (ed.) (1969). *The mathematical papers of Isaac Newton III, 1670-1673*. Cambridge University Press, Cambridge.

CHAPTER I

PLATO

§ 1. Plato's mathematical training

WHOEVER has written about the early history of mathematics has not failed to mention Plato, and usually in terms of praise. But the question of just how much he contributed to mathematical science is not easy to answer, and the anterior question of just how much mathematics he actually knew is still more difficult. There was plenty of mathematics to be known in his time. Heath is very definite on this point:

'There is therefore probably little in the whole compass of the Elements of Euclid, except the theory of proportion, due to Eudoxus, and its consequences, which was not in substance included in the recognized content of geometry in Plato's time.'[†]

Two comments suggest themselves. Another exception is the method of exhaustion, also due to Eudoxus, although some theorems first firmly established by this method were known before his time. But Eudoxus was early in life Plato's pupil, and it is hard to believe that these two monuments to Greek mathematical genius were unknown to the master. Whether he understood them completely is another question.

Plato certainly had ample opportunity to learn mathematics. Not from Socrates, who had little interest in the subject, but from Theodorus of Cyrene with whom he studied. He must also have been well acquainted with the mathematics of the Pythagoreans. He visited Sicily more than once, and he refers to the Pythagorean philosophy in various places. But our strongest testimony as to his mathematical knowledge is in the writing of that most laudatory commentator Proclus:

'But Plato, who was posterior to these, caused geometry as well as the other mathematical disciplines to receive a remarkable addition on account of the great study he bestowed on their investigation. This he himself manifests, and his books, replete with mathematical discoveries, evince.'[‡]

§ 2. Discussion of the value of mathematics

The most certain fact is that Plato had a very high opinion of the importance of mathematics and that, for various reasons. This appears in many places in his writings, but especially in *The Republic*, 525–38, from which I now quote at length, following Jowett's translation.

Socrates. And all arithmetic and calculation have to do with number?
Glaucon. Yes.
S. And they appear to lead the mind towards truth?
G. Yes, in a very remarkable manner.

[†] Heath (q.v.), vol. i, p. 217.　　　　　　[‡] Proclus (q.v.), book ii, p. 100.

S. Then this is the kind of knowledge for which we are seeking, having a double use, military and philosophical; for the man of war must learn the art of number or he will not know how to array his troops, and the philosopher also because he has to rise out of the sea of change, and lay hold on true being, and must therefore be an arithmetician.

G. That is true.

S. And our guardian is both warrior and philosopher?

G. Certainly.

S. Then this is the kind of knowledge which legislation may fully prescribe, and must endeavour to persuade those who are to be the principal men of our state to go and learn arithmetic, not as amateurs, but they must carry on the study till they see the nature of numbers with the mind only; nor again like merchants or retail traders with a view to buying or selling, but for the sake of military use, and for the soul itself and because this will be the easiest way for her to pass from becoming to truth and being.

G. That is excellent.

S. Yes, and now having spoken of it I must add how charming the scheme is and in how many ways it conduces to our desired end, if pursued in the spirit of a philosopher, and not of a shopkeeper!

G. How do you mean?

S. I mean, as I was saying, that arithmetic has a very great and elevating effect, compelling the soul to reason about abstract number, and rebelling against the introduction of visible or tangible objects into the argument. You know how steadily the masters of the art repel and ridicule anyone who attempts to divide absolute unity, when he is calculating, and if you divide they multiply, taking care that one shall continue one, and not become lost in fractions.

G. That is very true.

S. Now suppose a person were to say to them 'O my friends, what are these wonderful numbers, about which you are reasoning, in which, as you say, there is unity such as you demand, and each unit is equal, invariable, and indivisible', what would they answer?

G. They would answer, as I should conceive, that they were speaking of those numbers which can only be realized in thought.

S. Then you see that this knowledge can be called truly necessary, necessitating, as it clearly does, the use of pure intelligence in the attainment of pure truth.

G. Yes, that is a marked characteristic of it.

S. And have you further observed that those who have a natural talent for calculation are generally quick at every other kind of knowledge; and even the dull, if they have an arithmetical training, although they may derive no other advantage from it, always become much quicker than they would otherwise have been?

G. Very true.

S. And indeed you will not easily find a more difficult study, nor many as difficult.

G. You will not.

S. And for these reasons arithmetic is a kind of knowledge in which the best natures should be trained, and which must not be given up.

G. I agree.

S. Let this then be one of our subjects of education. And next shall we inquire whether the kindred science also concerns us ?

G. You mean geometry ?

S. Exactly so.

G. Clearly we are concerned with that part of geometry which relates to war, for in pitching a camp or taking a position or closing or extending the lines of an army, or any other military manœuvre, whether in actual battle or on the march, it will make all the difference whether a general is or is not a geometrician.

S. Yes, but for that purpose a very little either of geometry or calcula-tion will be enough, the question relates rather to the greater and more advanced part of geometry, whether that tends to make more easy the vision of the idea of the good, and thither, as I was saying, all things tend which compel the soul to turn her gaze towards that place where is the full perfection of being, which she ought, by all means, to behold.

G. True.

S. Then if geometry compels us to view being, it concerns us, if becoming only it does not concern us.

G. Yes, that we assert.

S. Yet anybody who has the least acquaintance with geometry will not deny that such a conception of the science is in flat contradiction to the ordinary language of the geometricians.

G. How so ?

S. They have in view practice only, and are always speaking in the narrow and ridiculous manner of squaring, and extending and applying and the like ; they confuse the necessities of geometry with those of daily life, whereas knowledge is the real object of the whole science.

G. Certainly.

S. Then must not a further admission be made ?

G. What admission ?

S. That the knowledge at which geometry aims is knowledge of the eternal and not of anything perishing or transient ?

G. That may readily be allowed and is true.

S. Then, my noble friend, geometry will draw the soul towards truth, and create the spirit of philosophy and raise up that which is now, unhappily, allowed to fall down.

G. Nothing will be more likely to have such an effect.

S. Then nothing should be more sternly laid down than that the inhabi-tants of your fair city should, by all means, learn geometry. Moreover, the science has indirect effects which are not small.

G. Of what kind ?

S. There are the military advantages of which you spoke, and in all departments of knowledge, as experience proves, anyone who has studied geometry is infinitely quicker of apprehension than one who has not.

G. Yes, indeed, there is an infinite difference between them.

S. Then shall we not propose this as second kind of knowledge which our youth must study ?

G. Let us do so.

It is perfectly clear from all this that Plato had a very exalted idea of the importance of mathematical study, finding three different reasons for it. First it is evident that mathematics has great practical utility. This surely could not be denied, at least as concerns certain parts of mathematics, but he considers it of minor importance. Socrates was little impressed by Glaucon's statement that a general needs a know-ledge of mathematics to set the battle in order. Secondly he believed that mathematics affords valuable mental training, for he asserts that anyone who has studied geometry is infinitely quicker of apprehension than one who has not. Perhaps the truth of this assumption is not decided yet. Our psychologists are by no means at one on the subject of the transfer of mental aptitudes, but Plato was troubled with no such question. And lastly he found the supreme justification of mathematics in this, that it leads to absolute truth, as it exists in the mind of God. The search for this is the highest object of man's endeavour.

§ 3. Analysis and synthesis

While we are occupied with Plato's views on the underlying philo-sophy of mathematics it is well to mention the subject of analysis. We have a first note on this subject in Proclus.† 'But there are, nevertheless, certain most excellent methods delivered, and one which reduces the thing sought by resolution to its explored principles which, as they say, Plato delivered to Leodamas, and from which he is reported to have been the inventor of many things in geometry.' We find a confirmation of this elsewhere. 'He was the first to explain to Leodamas of Thasos the method of analysis.'‡ And what, pray, is analysis ? The classical defini-tion is an interpolation in Euclid, Book XIII, Proposition 1. I follow Heath's translation, see his *Euclid*, vol. i, p. 138. Curiously enough, when Heath returns to this same in connexion with Book XIII he gives a different and less comprehensible translation.

'Analysis is an assumption of that which is sought, as if it were admitted, and the passage, through its consequences, to something which is admitted true. . . . Synthesis is an assumption of that which is admitted, and the passage through its consequences to the finishing or attainment of that which is sought.'

Now what all this means is the following. We wish to prove something. We begin by assuming it true. We deduce from this assumption a series

† Proclus (q.v.), book iii, p. 25.   ‡ Diogenes Laertius (q.v.), p. 299.

of consequences until we arrive at something which we recognize as valid. We call this analysis; synthesis is the reverse operation. We start with a statement which we recognize as correct and deduce therefrom a succession of consequences until we finally reach the thing which we wish to prove appearing as a necessary result of that which was admitted. This is especially clear if we use the modern mathematical jargon. In analysis we find a necessary condition for the truth of that which we wish to prove. In synthesis we show that this necessary condition is also sufficient.

Let me illustrate by a simple example. Suppose that we wish to show that the locus of all points in a plane equidistant from two given points is the perpendicular bisector of the line segment connecting them. Let the points be A and B, M the point midway between. If P be equidistant from them, the triangles $\triangle PMA$, $\triangle PMB$ are equal by three sides, the angles $\angle PMA$ and $\angle PMB$ are equal and supplementary, hence PM is the perpendicular bisector. Conversely, if P be on the perpendicular bisector, triangles $\triangle PMA$, $\triangle PMB$ are equal by two sides and the included angle, hence $PA = PB$.

But the question now arises, Was Plato really the discoverer of this method? It seems to me there is room for considerable doubt. The statement is open to discussion from two sides. If Plato invented the method, and communicated it to Leodamas, why do we not find it in precise form in Euclid? On the other hand, we do find it there in concealed form in various propositions which were surely known before Plato's time. The method of *reductio ad absurdum* consists in assuming, not the desired proposition, but the contrary, and showing that this leads to the contradiction of an admitted fact. I cannot believe that Plato really invented this method of attack. The most that I will concede is that he classified or popularized a procedure already in use. Heath dissents from this view, maintaining that the analysis and synthesis which Plato invented was dialectical, not mathematical:

'On the other hand, Proclus' language suggests that what he had in mind was the philosophical method described in the passage in the *Republic* which, of course, does not refer to mathematical analysis at all. It may well be true that the idea that he discovered the method of analysis is due to a misapprehension.'†

The passage in the *Republic* is a very obscure one which we shall come to presently, but I cannot get away from Proclus' statement that Leodamas discovered many things in geometry as a result of Plato's method of analysis; this must have been mathematical, not dialectical, analysis.

· † Heath (q.v.), vol. i, p. 291.

§ 4. Hypotheses

Anyone interested in the philosophy of mathematics will surely occupy himself with the hypotheses. It is true that the full significance of the modern conception of hypotheses is of recent growth—witness, for example, Poincaré's *Science et Hypothèse*—but we should expect something interesting on the subject from Plato; the result is most disappointing. We find an obscure passage which has worried commentators from the earliest times to our own day. It comes at the close of the *Republic*, Book vi. 510–11. Socrates has been expounding a very difficult idea, and Glaucon says that, frankly, he does not understand it.

Socrates. Then I will try again, you will understand me better when I make some preliminary remarks. You are aware that the students of geometry, arithmetic and kindred sciences, assume the odd and the even, and the figures, and the three kinds of angles, and the like in their several branches of science; these are the hypotheses which they and everybody are supposed to know, and therefore they do not deign to give any account of them, either to themselves, or to others; but they begin with them, and go on until they arrive at last in a consistent manner at their conclusion.

Glaucon. Yes, I know.

S. And do you not also know that though they make use of visible forms and reason about them, they are thinking, not of these, but of the ideals which they resemble, not of the figures which they draw, but of the absolute square and the absolute diameter and so on, the figures which they draw or make and which have shadows and reflections in water of their own, are converted by them into images, but they are really seeking to behold the things themselves, which can only be seen in the eye of the mind.

G. That is true.

S. And of this kind I spoke as intelligible although in the search after it the soul is compelled to use hypotheses not ascending to a first principle because she is unable to rise above the region of hypothesis, but employing the objects of which the shadows below are resemblances in their turn as images, they having in relation to the shadows and reflections of them greater distinctness and therefore higher value.

G. I understand that you are speaking of the province of geometry and the sister arts.

S. And when I speak of the division of the intelligible you will understand me to speak of the other sort of knowledge which reason itself attains by the power of dialectic, using the hypotheses, not as first principles, but only as hypotheses, that is to say, as steps and points of departure into a world which is above hypothesis, in order that she may soar above them to the first principle of the whole, and clinging to this, and then to that which depends on this, by successive steps she again ascends without the aid of any sensible object, from ideas, through ideas, and in ideas she ends.

These are surely dark sayings. Plato seems to recognize three objects of thought: the visible forms, the ideas which they represent, and the hypotheses, whatever these latter may be. He rightly states that the geometer is not dealing with the visible figures, that is to say with the drawings, but with the forms which they represent, and which are given by rigid mathematical definition. But what are the hypotheses ? To us a mathematical hypothesis is an assumption that something exists or something is true. This does not fit into Plato's scheme at all. His first idea seems to be that an hypothesis is an existence assumption for such things as odd or even, the three kinds of angles, etc. But what then does he mean by saying that the soul is unable to rise above hypotheses to first principles ?

Persons adept in explaining Plato have been not a little puzzled by this passage which is so incomprehensible to a mere mathematician. I quote Adam :

'It appears therefore that the ὑπόθεσες of dialectic are not like those of mathematics, immoveable and fixed, and that we may be called upon to render an account of them, nay more, that it is our duty to submit them to examination ourselves.'†

Perhaps, in the *Republic*, Plato is speaking of the hypotheses of dialectic, for he looks upon dialectic as a higher approach to truth than mathematics ; we cannot argue the point with him, certainly nothing could exceed the care with which twentieth-century mathematicians have examined their hypotheses. I find somewhat more help in the opinion of Jowett.

'There is a truth, one and self-existent, to which by the help of a ladder let down from above, the human intelligence ascends. . . . It is the idea of the good, and the steps of the ladder leading up to the highest or universal existence are the mathematical sciences, which also contain in themselves elements of the universal. These, too, are seen in a new manner when we connect them with the idea of the good. Then they cease to be hypotheses or pictures and become parts of the higher truth which is at once their first principle and their final cause.'‡

I think all this means that essential truth exists somewhere outside of ourselves, say in the mind of God, that mathematics is a ladder by which we may ascend towards it, that owing to the frailty of our natures we cannot do so without first making assumptions, but that the truth of the final conclusion does not depend on the validity of these assumptions but is independent of ourselves. But it may all mean something quite different.

§ 5. Definitions

We should expect that Plato would pay attention, not only to the hypotheses of mathematics, but to some at least, of the definitions. Here

† Plato², vol. ii, p. 175.   ‡ Plato¹, pp. xcv, xcvi.

he is strangely remiss, as it seems to me. In the *Meno* 75 and 76 is a discussion of what a figure is. Socrates proposes this definition: 'A figure is the only thing which always follows colour.' We should say that such a statement makes no sense whatever, but Meno objects that a person might say that he does not know what colour is any more than he knows what a figure is. Socrates continues:

Socrates. You will acknowledge that there is such a thing as an end or termination or extremity. All such words I use in the same sense, although I am aware that Proclus might draw distinctions about them, but still you, I am sure, would speak of a thing as ended or terminated, that is all which I am saying, not anything very difficult.

Meno. Yes I should, and I believe that I understand your meaning.

S. And would you speak of a surface and also of a solid as, for example, in geometry?

M. Yes.

S. Well then you are now in a condition to understand my definition of a figure. I define a figure to be that in which a solid ends, or, more concisely, as the limit of a solid.

It appears from this that a figure is a surface, which is rather strange in that he has recently spoken of a surface. Moreover, at the end of *Meno* 74 he spoke of a figure as having roundness or straightness. Presumably he means that a figure is a termination. The termination of a surface is one-dimensional and has roundness or straightness, but the termination of a solid is a surface. Why this should be said to come after colour I cannot imagine.

We find somewhat similar ideas in *Parmenides* 137 where Plato is seeking to define the 'one' or unity, and establishing as many contradictions as possible in this idea. It cannot have parts or be surrounded, it can never move or be at rest. But at the end of 137 we have these interesting definitions:

Parmenides. The one having neither beginning nor ends is unlimited.

Aristoteles. Yes, unlimited.

P. And therefore formless, it cannot partake of either the round or the straight.

A. But why?

P. Why the round is that of which all the extreme points are equidistant from the centre.

A. Yes.

P. And the straight is that of which the centre intercepts the view of the extremes.

A. Yes.

It is interesting to compare this with Euclid's definitions: 'A circle is a plane figure contained by one line such that all straight lines falling upon it from one point lying within the figure are equal. . . . A straight line is a line which lies evenly within the points on itself.'

It is worth noting the fundamental difference between these two definitions. The first is pragmatic; it gives that property of the circle which is the basis of all that we know about the figure; the fact that a straight line lies evenly is not involved in any theorem having to do with a straight line.

§ 6. The Pythagorean theorem

It is time to turn from what Plato says about mathematics philosophically considered, to the few strictly mathematical passages which occur in his works. I take the geometrical ones first, then the arithmetical ones. Plato certainly was familiar with the Pythagorean theorem, but the only direct mention of it is in the special case where we have an isosceles right triangle. This appears at length in *Meno* 82–4. Socrates is trying to prove our pre-existence by showing how, through skilful questioning, a person can come to realize a truth which is not actually stated to him, but must come as a recollection of something learnt in an earlier state. He calls up a Greek boy who has had no previous training in geometry, and asks him if he knows what a square is. The boy replies in the affirmative. Socrates then draws a square, divides it into four equal parts by connecting the mid-points of the opposite sides. We have a square, two units on each side, with an area of four units. The boy acknowledges this, and that a square of double the size would contain eight. Socrates asks him to construct such a square. He doubles the dimensions and finds that the area is not eight but sixteen. The boy is given a second chance. He sees that the side must be greater than two but less than four; he tries three. This time the area is not eight but nine, not right but better.

Socrates then makes suggestions. The large square is made up of four little squares. If we connect the middle points, not of the opposite sides, but of the successive sides, we find that we cut off half the area of each of the four little squares so that we have a square of area two, half of the original area. Its sides are the diagonals of the small squares, and we see that the square on the diagonals is the sum of the squares on the sides.

Socrates. Without anyone teaching him he will recover this knowledge for himself, if he is asked questions?
Meno. Yes.
Soc. And this spontaneous recovery is recollection?
Meno. True.

I am afraid that nowadays no one would agree with this conclusion. The conviction comes from looking at a very simple figure where areas can be estimated immediately. Plato would doubtless have liked to prove the general Pythagorean theorem in this fashion, but that was impossible.

This simple piece of geometry is followed by a problem which has intrigued the critics a good deal; it comes in *Meno* 87.

'At any rate will you condescend a little and allow the question "whether virtue is given by intuition, or in some other way", to be argued by hypothesis. As the geometrician, when he is asked whether a certain triangle is capable of being inscribed in a certain circle will reply, "I cannot tell you as yet, but I will offer a hypothesis which will assist us in forming a conclusion."

If the figure be such that when you have produced a given side of it the given area of the triangle falls short by an area corresponding to the point produced, then one consequence follows, and if this is impossible, then some other. And therefore I wish to assume a hypothesis before I tell you whether this triangle is capable of being inscribed in the circle.'

This is Jowett's translation and is as blind as one could ask for. Heath does somewhat better:

'If the given area is such that when one has applied it (as a rectangle) to the given straight line in the circle it is deficient by a figure (rectangle) similar to the very figure which is applied, then one alternative seems to me to result, while again another results when it is impossible for what I have said to be done to it.'†

The one certain thing here is that Plato was familiar with the application of areas, the Greek method of solving mixed quadratic equations which was discovered by the Pythagoreans. Heath tells us‡ that by 1861 thirty different explanations of the puzzle had been published, and doubtless others since. I will limit myself to two. First of all Plato probably realized that the maximum triangle inscriptible in any circle is equilateral, so that if it was a question of whether a given triangle was greater than the equilateral triangle inscribable in a given circle, this certainly should not have been beyond the Greek geometry of his time. But when it comes to actually constructing the triangle in a given case, the matter is not so simple. Heath says that the most popular solution is that of Benecke (q.v.). This writer points out that Plato has just been talking of isosceles triangles, and assumes that he continues to do so. An isosceles right triangle will fit into a circle if its hypotenuse be equal to the diameter. In that case the area of the triangle is that of a square on a radius, which has the same shape as the square on the other half of that diameter. I agree with Heath in thinking that this is too simple. Another explanation is that of August. I have not seen the original of this, but it is apparently reproduced in Butcher (q.v.). Let us limit ourselves to the isosceles triangle. If an isosceles triangle be

† Heath (q.v.), vol. i, p. 299.   ‡ Ibid., p. 298.

inscriptible in a given circle, the bisector of the vertical angle will be a diameter (the line within the circle). The area of the triangle is, then, that of a rectangle of which one side is the altitude lying along the diameter, and this is similar to the rectangle on the remainder of the diameter and the other half of the base when, and only when, the triangle is inscriptible. Benecke objects to this on philological grounds which I am unable to follow. I merely remark that Butcher was a very distinguished scholar unlikely to go far wrong in such a matter. I am bound to confess that the whole thing seems to me rather a futile speculation. Plato must have been familiar with Euclid, IV. 5, the problem of circumscribing a circle about a given triangle; all he had to do was to compare the radii of the two circles.

§ 7. Two mean proportionals

We come to another more interesting speculation about Plato's geometry in Eutocius' commentary on Book II of Archimedes' work on the sphere and cylinder. It will be found in Archimedes,† and is part of a most important passage dealing with the history of Greek mathematics. We have here a discussion of the various methods suggested for inscribing two mean proportionals between two given lengths, or numbers. No less than thirteen of these are offered, that described as *ut Plato* is like this (Fig. 1).

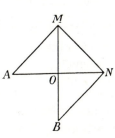

Let AO and OB be the two given lengths OM, ON, the two mean proportionals.

$$\frac{AO}{OM} = \frac{OM}{ON} = \frac{ON}{OB}.$$

Fig. 1.

We seek the points M and N. Clearly $\angle AMN$, $\angle MNB$ are right angles. We take an instrument like a vice, or the simple arrangement which shoe-makers use to determine the length of the human foot. Essentially we have two parallel jaws which move in such a way that all points trace straight lines perpendicular to the two. We open them so that A lies on one jaw, B on the other, then twist and regulate the opening until a pair of corresponding points lie on AO and OB respectively.

This is surely ingenious, but difficult to accomplish with any degree of accuracy; the important question is 'Did Plato really discover or propose this?' Eutocius is a responsible writer, one should not lightly set his verdict aside. But it is strangely out of keeping with our ideas about Plato. We have seen how he laid down the axiom that geometry aims at the knowledge of the eternal, not of anything transitory or perishing. Plutarch puts the matter in the strongest fashion.

† Vol. iii, pp. 67 ff.

'Eudoxus and Archytas had been the first originators of this far famed and highly prized art of mechanics which they employed as an elegant illustration of geometrical truths, and as a means of sustaining experimentally to the satisfaction of the senses, conclusions too intricate for proofs by words or diagrams. As, for example, to solve the problem, so often required in constructing geometrical figures, given two extremes, to find two mean lines of proportion: both these mathematicians had recourse to the aid of instruments adapting to their purposes certain curves, and sections of lines. But what with Plato's indignation at it, and his invectives against it as the mere corruption and annihilation of the good of geometry, which was thus turning its back on the unembodied objects of pure intelligence, to recur to sensation, and ask help (not obtained without base subservience and deprivation) from matter, so it was that mechanics came to be separated from geometry and repudiated and neglected by philosophers, and took its place as a military art.'†

This seems to come much closer to the idea which we get of Plato from other passages. The two can only be reconciled if we imagine that Plato invented the mechanical method of inserting two mean proportionals, only to reject it as unworthy.

§ 8. Geometry in three dimensions; regular solids

Plato's interest in geometry was not confined to the plane but extended to geometry in space, although he was pessimistic about the state of this subject. In the continuation of a passage, already quoted, from the *Republic*, he apologizes for failing to mention solid geometry when discussing the important subjects for instruction, but passing directly to astronomy. We find in the *Republic* 528:

Glaucon. Yes, there is a remarkable charm in them. But I do not clearly understand the change in the order. First you begin with a geometry of plane surfaces.

Socrates. Yes.

G. And you placed astronomy next, and then you made a step backwards ?

S. Yes, and I have delayed you by my hurry, the ludicrous state of solid geometry, which in natural order should have followed, made me pass on and go over to astronomy or motion of solids.

Plato was particularly interested in the five regular solids, but in their mystical rather than their mathematical properties. They were well known in his time, having been discovered by the Pythagorean School. *Here Proclus is our informant:

'But after them the Pythagoreans changed the philosophy, which is

† Plutarch (q.v.), vol. ii, pp. 252, 253.

* This translation from Proclus has drawn much criticism, for example from Burkert ([1972], 409-11). In the translation given by Morrow it runs: 'Following upon these men, Pythagoras transformed mathematical philosophy into a scheme of liberal education, surveying its principles from the highest downwards and investigating its theorems in an immaterial and intellectual manner. He it was who discovered the doctrine of proportionals and the structure of the cosmic figures.' (Proclus [1970], 52-3.)

conversant about geometry into the form of a liberal doctrine, assuming its principles in a more exalted manner, and investigating its theorems immaterially and intellectually; who likewise invented a treatise on such things as cannot be explained in geometry, and discovered the construction of mundane figures.'

It would seem that anyone who wrote about such things as cannot be explained in geometry would have pretty much the whole intellectual field open to him. I presume what is meant is geometrical figures which are not contained in the plane. As for the identification of the regular solids with 'mundane figures', we find the explanation in *Timaeus* 53 ff.

'In the first place, then, as is evident to all, fire and earth, water and air are bodies. And every sort of body possesses solidity, and every solid must be contained in planes, and every plane rectangular figure is composed of triangles.... Thus proceeding by a combination of probability with demonstration, we assume them to be the original elements of fire and the other bodies.'

Plato then proceeds to build the World out of triangles, in fact out of right triangles, as every triangle can be divided into two right triangles. Of the right triangles he picks two as being the most interesting, the isosceles, which are halves of squares, and the thirty-sixty-ninety ones, which are halves of equilateral ones. He takes six of these latter and makes an equilateral triangle out of them; why two would not have done just as well I do not know. Out of four equilateral triangles he makes a regular tetrahedron. This is the simplest, and so the most fundamental of the regular solids, and he associates it with fire, the most fundamental of the elements. With eight equilateral triangles we can construct a regular octahedron, which is assigned to air; twenty equilateral triangles will give the regular icosahedron which is assigned to water. These are the only regular solids made from the thirty-sixty-ninety triangles; two isosceles will give a square and six squares will give a cube, a hard resistant body assigned to earth. There remains the dodecahedron, which he dismisses casually: 'There was yet a fifth combination which God used in the delineation of the Earth.'

These ideas are surely pre-Platonic. Four regular solids were discovered. These were believed to be the only ones, they fitted in nicely with the four supposed elements. It must have been awkward when a fifth regular solid was discovered, with no element to correspond; the assignment to the whole Creation is weak enough. I seize especially on the word 'delineate', for this emphasizes the fact that the association must be understood in a symbolical sense. Plato knew enough about combustion to know that fire was not composed of tetrahedra; the tetrahedron is a symbol of the idea of fire as it appears to the mind of God.

Besides the regular solids Plato seems to have been interested in what are called semi-regular solids; the vertices lie on a sphere, the faces are regular polygons, equal in groups. Here is Heiberg's translation of what Hero of Alexandria has to say on the subject:

'Euklides hat im XIII. Buch der Elemente (13–17) bewiesen, wie er diese fünf Körper mit einer Kugel umfasst, er nimmt nämlich nur die Platonischen an. Archimedes aber sagt, es gäbe, im ganzen, dreizehn Körper, die in einer Kugel eingeschrieben werden können, indem er ausser den gennanten fünf noch acht hinzugefügt, von diesen habe auch Plato das Tessareskaidekaeder gekannt; dies aber sei ein zweifaches, das ein, aus acht Dreiecken und sechs Quadraten . . . das andere umgekehrt aus acht Quadraten und sechs Dreiecken, welches schwieriger zu sein scheint.'†

The adjective *schwieriger* is an under-statement; the solid does not exist, as we can prove by applying Euler's formula for faces, edges, and vertices.

§ 9. Numbers and commensurability

Plato was very much interested in numbers, as were the Pythagoreans, who believed that numbers were the basis of everything. In the *Republic*, 546, he speaks of a geometrical number 'which represents a geometrical figure which has control over the good and the evil of births'. Such a statement shows how essentially mystical was the whole idea. It does not seem to me worth while to go into the various surmises which have been made as to what is really meant, merely referring to Heath.‡

Plato was fascinated by the idea of incommensurable lengths and irrational numbers; we have an interesting passage in the *Laws* 819–20:

Athenian Stranger. And again in the measurements of things which have length and breadth and depth they free us from the natural ignorance of these things which is so ludicrous and disgraceful.

Cleinias. What kind of ignorance do you mean?

A. O my dear Cleinias, I, like yourself, have late in life heard with amazement of our ignorance of these matters; to me we appear to be more like pigs than men, and I am quite ashamed, not only of myself, but of all Hellenes.

C. About what? Say stranger, what do you mean?

A. I will, or rather, I will show you my meaning by a question, and do you please to answer me. You know, I suppose, what length is?

C. Certainly.

A. And what breadth is?

C. To be sure.

A. And you know that these are two distinct things, and that there is a third called depth?

C. Of course.

---

† Hero (q.v.), vol. iv, pp. 65, 67.   ‡ q.v., vol. i, pp. 305 ff.

A. And do not all of these seem to you to be commensurable with themselves ?

C. Certainly.

A. That is to say, length is naturally commensurable with length, and breadth with breadth, and depth, in like manner, with depth ?

C. Undoubtedly.

A. But if some things are commensurable, and others wholly incommensurable, and you think that all things are commensurable, what is your position with regard to them ?

C. Certainly far from good.

A. Concerning length and breadth when compared with depth, or breadth and length when compared with one another, are not all Hellenes agreed that they are commensurable with one another in some way ?

C. Quite true.

A. But if they are absolutely incommensurable, and yet all of us regard them as commensurable, have we not reason to be ashamed of our compatriots ?

This as it stands makes very little sense. I think that Plato must have meant that anyone would naturally assume that the length and breadth of the same rectangle must be commensurable, yet this is not necessarily the case. The breadth of a rectangle might be that of a square, while its length is equal to the diagonal of the square.

Plato was fond of certain individual numbers. In the *Laws* he shows a partiality for 12. 'There is no difficulty in perceiving that the twelve parts admit of the greatest number of divisions of that which they include, or in seeing that the other numbers which are consequent upon them and are produced out of them up to 5,040.'

What interests him is to find which numbers have the greatest number of subdivisions into equal parts. Thus 12 can be divided into 2, 3, 4, or 6 equal parts. As for 5,040, which is 7!, that can be subdivided in a fantastic number of ways. Plato closes with this cheerful remark: 'Above all arithmetic stirs up him who is by nature sleepy and dull and makes him quick to learn, retentive, shrewd and aided by the art divine, makes progress beyond his natural powers.'

There is a curious note in *Timaeus* 321 where he says: 'Two terms must be united by a third which is a mean between them, and had the Earth been a surface, only one term had sufficed, but two terms are needed with solid bodies.' This apparently means that if two integers are perfect squares, the mean proportional between them is an integer, but if we have two integers which are perfect cubes, there are two integral mean proportionals.

I said above that Plato was interested in the Pythagorean theorem. This interest extended to the formation of integral right triangles, that is to say, to finding pairs of integers, such that the sum of their squares is the square of an integer. I have not seen this in Plato's own work. We read in Proclus, after an account of a Pythagorean method based on pairs of odd numbers:

'But the Platonic method originates from even numbers. For when he has assumed an even number he places it as one of the sides about a right angle, and when he has divided it in halves, and has produced a quadrangular number from the half when he adds unity to this he forms the subtending side (hypotenuse), but when he has taken unity from the quadrangle he forms the remaining side about the right angle.'†

The numbers are $2n$, n^2+1, and n^2-1.

I mentioned quadrangular numbers. Plato, and after him Euclid, was interested in cataloguing numbers according to such distinctions. I shall return to this point presently. He was constantly preoccupied with irrationals, which he ordinarily reached through the diagonals of squares, but he was aware of the existence of other irrationals. Thus we find in *Hippias Major* 303:

'Of what kind, then, Hippias, does the beautiful seem to you? Whether as you asserted, that if you and I are strong, both are so, and if both so, then is each; and similarly if I and you are beautiful, both are so, and if both so too is each. Or is there nothing to prevent it as (in the case of numbers where) some things taken together being even, may be when taken singly odd or even, or when each is taken separately is irrational, but taken both together may be rational or, perhaps, irrational?'‡

This shows that Plato knew that the sum of two irrationals might be rational, which amounts to saying that not all irrationals are square roots of integers, but might be mixtures of integers and square roots. The most famous passage in which Plato introduces irrationals is *Theaetetus* 147:

Theaetetus. Yes, Socrates, there is no difficulty as you put the question. You mean, if I am not mistaken, something like what occurred to me and to my friend here, your namesake Socrates, in recent discussions.

Socrates. What was that, Theaetetus?

T. Theodorus was writing out for us something about roots, such as roots of three or five, showing that in linear measure they are incommensurable by the unit; he selected other numbers up to seventeen—there he stopped. Now as there are innumerable ratios, the notion occurred to us to include them all under one name or class.

S. And did you find such a class?

T. I think we did, and I should like to have your opinion.

S. Let me hear.

---

† Proclus (q.v.), p. 203.

‡ *Hippias Major* does not appear in Jowett. I have followed Burge's *Plato*, London, 1755, p. 256.

T. We divided all numbers into two classes, those which are made up of two equal factors multiplying into one another, which we represented as square or equilateral numbers, that was one class.

S. Very good.

T. The intermediate numbers as three, five, and every other number which is made up of unequal factors, either a greater multiplied by a less or a less multiplied by a greater, and when regarded as a figure is contained in unequal sides, all these we represented as oblong numbers.

S. Capital, and what follows ?

T. The lines or sides which have for their squares the equilateral plane numbers are called by us lengths or magnitudes, and the lines which are the roots of (or whose squares are equal to) oblong numbers were called powers or roots ; the reason of this latter name being that they are commensurable with the others (i.e. with the so-called lengths or magnitudes) not in linear measurement, but in the value of their squares.

Heath says† that this passage has given rise to various conjectures as to whether Theodorus had some way of approximating to square roots. I confess that I see nothing of the sort, nor do I believe that Plato would have been much interested in any method of approximation to the value of irrationals. To me it seems that Theodorus, recognizing integers either as perfect squares or oblong, showed that the square root of an oblong number is irrational, following the classical proceeding for the square root of 2. The objection that this is too simple to have excited Plato's admiration seems to me ill founded. The terms 'commensurable' and 'incommensurable in square' appear constantly in Euclid X. Two small comments also occur to me. Is it possible that Plato saw a difference between multiplying a less by a greater, and a greater by a less ? Or again, the number 6 would seem to be a length or magnitude, as it is the root of 36, but it is thus the root of 9×4, and so should be called a power or root. It is passing strange that he does not draw the distinction between prime and not prime numbers.

And what shall be our final judgement of Plato as a mathematician ? A productive mathematical scholar he certainly was not. He had the highest opinion of the value, especially the spiritual value, of the subject. He had a bowing acquaintance with all of the pure mathematics of his time. He frequently, I almost said habitually, expressed himself obscurely if not inaccurately. The vital question seems to me to be, Is this accident or design. Did he really have a true grasp of the subjects he touched upon, or had he merely heard them spoken of, never really grasping their significance ? The passage where he brings in the application of areas suggests a real knowledge, but all of the rigmarole about hypotheses is discouraging. Did he have a real knowledge, but confine his interest to the mystical side, or had he really failed to penetrate the significance of

† Vol. i, pp. 204–6.

the great triumphs of Greek mathematics which were known in his time ? Each must decide for himself. I must confess to being enough of a Philistine to believe it to be a waste of time to look for important facts of a mathematical or historical nature under the confused statements which may indeed arise from the spiritual height of his views, but equally from a fundamental lack of grasp in his knowledge.

CHAPTER II

OMAR KHAYYÁM

§ 1. The story of Omar

GHIZATHUDIN ABULFATH IBN IBRAHIM AL KHAYYAMI, who is generally known under the name given at the beginning of this chapter, spent most of his life at Naishápúr. There is a good deal of uncertainty about his dates, but he seems to have lived during the latter part of the eleventh and the beginning of the twelfth centuries. There is a tradition that he died in 1123, but we cannot place too much confidence in it. The most significant episode in his early years was his friendship with two unusual young men, Nizam al Mulk and Hassan ibn Subbuh. According to the legend these boys agreed that if one of them should come to fame and fortune he would show kindness to the other two. The lucky one was Nizam, and he undertook to carry out the promise. He made Hassan Court Chamberlain. This was a poor move. Hassan turned out to be a troublesome courtier, and was exiled from the Court. He became the head of an exceedingly blood-thirsty and troublesome band of fanatics called Ismailians. They seem to have specialized in assassination; some etymologists tend to derive this word from Hassan, but others connect it with hashish. Omar did not ask for anything so spectacular, he merely desired to be raised so far above want that he could give his life to his favourite studies, mathematics and astronomy. This modest request was granted; he made some return by his work in reforming the calendar.

Omar's fame as a scientist has, in recent years, been completely obliterated by his brilliant reputation as a poet. A good share of the credit for this belongs to his peerless translator, Edward Fitzgerald. I have no competence to express an appreciation here, neither is there any reason for me to discuss his anti-religious philosophy. Some persons have maintained that he was grossly immoral, a libertine addicted to unnatural vice. Perhaps he was, perhaps not. The impression which I get from reading the *Rubá'iyyat* is that of a sophisticated and disillusioned, but not unkindly cynic, who praises the attainable delights of the senses, and treats his adversaries with caustic wit. Very likely he was an atheist, but he was willing enough to use pious phrases of a conventional pattern. Here are the opening lines of Omar (q.v., Woepcke's translation): 'Au nom de Dieu, clément et miséricordieux. Louange à Dieu, seigneur des mondes, une fin heureuse à ceux qui le craignent, et point d'inimitié si ce n'est que contre les injustes. Que la bénédiction divine repose sur les prophètes, et particulièrement sur Mohammed et sur toute sa famille.' He closes his essay with these words: 'C'est Dieu qui facilite la solution de ces difficultés par ses bienfaits et sa générosité.' He frequently wishes that God shall be merciful to this or that other scientist. Such piety is common enough in Islamic writing; very likely Omar had his tongue in his cheek while expressing himself in this fashion, but these phrases did flow from his pen.

Omar wrote a treatise, now completely lost, which seems to have contained something of great interest in the history of mathematics. He writes: 'J'ai enseigné à trouver les côtés du carré-carré, du cube-cube etc. à une étendue quelconque, ce qu'on n'avait pas fait précédemment. Les démonstrations que j'ai données à cette occasion, ne sont que des démonstrations arithmétiques, fondées sur les parties arithmétiques des éléments d'Euclide.'†

what?

Tropfke expresses the opinion: 'Die letzte Bemerkung kann man offenbar nur auf Benutzung der binomschen Entwickelung auf beliebig hohe Exponenten deuten, wodurch dann Alkhajjami als Entdecker des Binomialtheorems für ganzzählige Exponenten anzusehen wäre.'‡

§ 2. Early study of cubic equations

Omar's mathematical interest is primarily centred in the solution of equations. Right at the start he sees two problems, what he calls the algebraic and the geometrical solution. He means by an algebraic solution one in positive integers. In all such questions the influence of Diophantus is paramount. Omar's work shows no great progress beyond what was attained by this master. He means by a geometrical solution one in terms of what he calls measurable quantities, lengths, areas, and volumes. Here is the statement of a typical problem: 'Un cube, des carrés et des nombres sont égaux à des côtés.'§ This means that we are concerned with such an equation as

$$x^3 + cx^2 + a^3 = b^2 x.$$

I have written this in homogeneous form, and shall continue to do so. The word *des* suggests that his coefficients were positive integers, but I am not sure that this is what he meant. A Greek would probably have said: 'We wish to find the line which has the property that a cube with this as an edge added to a square based rectangular parallelepiped of given height with this as a base edge, and added to a given volume, is equivalent to a rectangular parallelepiped of given square base with this as height'. This would be a more accurate statement, but Omar did not phrase things so accurately; his unknown is a length, determined by the way that it enters into certain areas and volumes. His great task is to

† Omar (q.v.), p. 13.   ‡ Tropfke (q.v.), 3rd ed., vol. ii, p. 174.
§ Omar Khayyám (q.v.), p. 47.

make a systematic study of all linear, quadratic, and cubic equations, which have at least one positive root. His method consists in using, not only what the Greeks called plane loci, straight lines and circles, but solid loci, conic sections. Before actually giving any of his work it will be well to indicate some of what had been accomplished by his predecessors; the solution of equations, especially cubic equations, had greatly intrigued the Arabian and Persian mathematicians.

The conic sections first appear in connexion with the problem of inserting two mean proportionals between given lengths. The traditional discoverer of this procedure was Menaechmus, a pupil of Plato's, who flourished in the fourth century B.C. If the given lengths be a and b, and the two means we seek x and y,

$$a/x = x/y = y/b.$$

From these we get two parabolas

$$x^2 = ay; \qquad y^2 = bx,$$

and the hyperbola $\qquad\qquad xy = ab.$

We have also the cubic equations

$$x^3 = a^2b; \qquad y^3 = ab^2.$$

We find the first extension of this method of treating cubics in Archimedes' treatise on 'The Sphere and Cylinder' in the fourth problem of Book II: 'To cut a given sphere by a plane so that the volumes of the segments bear to one another a given ratio'.† Archimedes did not give the solution in this place, but his commentator Eutocius found elsewhere a fragment which had the earmarks of being the work of the master; it is reproduced at length in the place just cited. I will greatly shorten the work by using modern algebraic notation. The radius of the sphere shall be r, the depth of the segment x. By proposition 2 of the same book the volume of the segment is

$$V = \frac{\pi x^2}{3}(3r-x).$$

Let us put this into words. We extend the diameter perpendicular to the base of the segment one radius length beyond that end which is away from the segment; we have a line (in the Greek sense) of length $3r$ divided into two segments of lengths x, $3r-x$. If the ratio of the two segments be m/n,

$$\frac{3V}{4\pi r^3} = \frac{m}{m+n}; \qquad \frac{x^2}{4r^2} = \frac{mr}{(m+n)(3r-x)}.$$

† Archimedes², pp. 62 ff.

Archimedes generalizes this into the problem of dividing a given line into two such parts that the ratio of one to a given length is the inverse of the ratio of the square of the other to a given area

$$\frac{x^2}{b^2} = \frac{a}{c-x}; \qquad x^3 + ab^2 = cx^2.$$

We seek the intersections of

$$x^2 = \frac{b^2}{c}y; \qquad y(c-x) = ac.$$

Here we have a parabola whose axis is parallel to one asymptote of a given hyperbola, so that we have two conics with only three finite intersections, and so are led to a cubic equation. In this fragment Archimedes gives a careful discussion of the limits on the coefficients necessary to ensure a positive solution. In the same place we find solutions by Dionysodorus and Diocles. The Arabs took much interest in the same problem. We find in Omar:[†]

'Quant aux modernes, c'est Almâhâni qui parmi eux conçut l'idée de résoudre algébriquement le théorème auxiliaire employé par Archimède dans la quatrième proposition de son traité de la sphère et du cylindre où il fut conduit à une équation renfermant des cubes, des carrés et des nombres, qu'il ne réussit pas à résoudre après en avoir fait une longue méditation. On déclare donc que cette résolution était impossible jusqu'à ce que parût Abou Djafar Alkhâzin (A.D. 960), qui résolut l'équation à l'aide des sections coniques.'

Another famous problem was that of trisection of the general angle. If we call this angle 3ϕ we have

$$4\cos^3\phi - 3\cos\phi = \cos 3\phi,$$

which leads to the cubic equation

$$x^3 = b^2x + a^3.$$

Early writers did not have a sufficient knowledge of trigonometry to lead to this simple equation. The Greeks preferred the use of the 'quadratic' for this purpose.[‡] Two good solutions appear in the work of Pappus.[§] The first of these calls for a preliminary construction. Given a rectangle $AB\Gamma\Delta$ (Fig. 2) to find E on $\Gamma\Delta$, Z on $B\Gamma$ so that $EZ = c$, a known length.

Let $A\Delta = a, \qquad AB = b.$

I seek H below Z so that $E\Delta HZ$ is a parallelogram.

† q.v., p. 2.   ‡ Archimedes[1], p. cxxxviii and Cantor[2], p. 196.
§ Pappus (q.v.), vol. i, pp. 212 ff.

Let H have the coordinates (x, y), the axes being ΔK, $\Delta \Gamma$

$$\Delta H^2 = x^2 + y^2 = c^2.$$

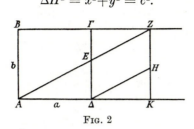

Fig. 2

From similar triangles

$$\frac{x}{a+x} = \frac{y}{b}; \qquad xy + ay = bx.$$

It is true that the two conics lead to a quartic and not to a cubic equation, but they give a solution of our problem of trisection. Let $\angle \Delta A\Gamma$ (Fig. 3) be the angle we wish to trisect. Construct the rectangle

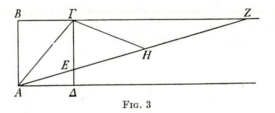

Fig. 3

$\triangle AB\Gamma$ and take $EZ = 2A\Gamma$ by the preceding construction, while H is the middle point of EZ

$$A\Gamma = EH = HZ = H\Gamma$$
$$\angle ZA\Gamma = \angle AH\Gamma = 2\angle HZ\Gamma = 2\angle \Delta AH$$
$$\angle \Delta A\Gamma = 3\angle \Delta AH.$$

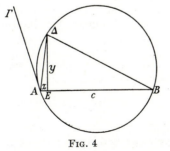

Fig. 4

Here is Pappus' other solution. Let $\angle \Gamma AB$ (Fig. 4) be the angle we wish to trisect. Let $A\Delta$ be a trisector. Pass a circle through BA tangent to $A\Gamma$:

$$\angle \Delta AB = 2\angle \Gamma A\Delta = 2\angle \Delta BA.$$

We seek the locus of Δ. Let $AE = x$, $\Delta E = y$,

$$\frac{y}{x} = \tan\angle\Delta AB; \qquad \frac{y}{c-x} = \tan\angle\Delta BA,$$

$$\frac{y}{x} = \frac{\dfrac{2y}{c-x}}{1-\dfrac{y^2}{(c-x)^2}},$$

$$2x(c-x) = (c-x)^2 - y^2.$$

We need, then, to find the intersection of a circle with a rectangular hyperbola.

The Arabs were very fond of the trisection problem; we find in Omar[†] some half-dozen examples of their work. Here is an Arab problem similar to Archimedes' problem of cutting the sphere. It is due to Al Kuhi (A.D. 1000) and given by Omar, p. 104. To construct a segment of a sphere which has a given volume, and also a given curved surface. The two unknowns shall be r the radius of the sphere, and x the depth of the segment. We have the equations

$$rx = b^2, \qquad x^2(3r-x) = a^3.$$

These lead to

$$3b^2x = x^3 + a^3; \qquad r^3 + \frac{b^6}{a^3} = 3\frac{b^4}{a^3}r^2.$$

Since

$$r > 0, \qquad x < 2r,$$

$$b < r\sqrt{2}; \qquad r^2 > \frac{b^2}{3},$$

$$\frac{9b^8}{a^6} > r^2 > \frac{b^2}{3}; \qquad 27b^6 > a^6.$$

We shall return presently to Omar and the solution of the cubic, but I must first mention an interesting opinion of Heath's:

'There can be no doubt that Archimedes solved this equation as well as the similar one with the negative sign, i.e. he solved the two equations

$$x^3 \pm ax^2 \pm b^2c = 0,$$

obtaining all their positive roots. In other words he solved completely, as far as the real roots are concerned, a cubic equation in which the term in x is absent, although the determination of the positive and negative roots of the same equation meant for him two separate problems. And it is clear that all cubic equations can be reduced to the type Archimedes solved.'[‡]

† q.v., pp. 117 ff.   ‡ *Archimedes*, cxxix.

It seems to me that here Heath is going beyond the facts. He says:[†] 'Though Archimedes does not give the solutions, the following considerations may satisfy us as to his method'. The method suggested, but not carried out, is based on Archimedes' work on spheroids and conoids. Very likely he could have carried it out, but he did not. And what reason is there to believe that Archimedes, or any ancient mathematician, could transform any cubic equation into one which lacks the term in x?

§ 3. Omar's own contribution

Omar's fame as a mathematician rests chiefly on his claim to be the first to handle every type of cubic that has a positive root. Here his claim is perfectly definite. He says:[‡]

'Après lui tous les géomètres avaient besoin d'un certain nombre des susdits théorèmes, et l'un en résolut un et l'autre un autre. Mais aucun d'eux n'a rien émis sur l'énumération des cas de chaque espèce, ni sur leur démonstration, si ce n'est relativement à deux espèces que je ne manquerai pas de faire remarquer. Moi, au contraire, je n'ai jamais cessé de désirer vivement de faire connaître avec exactitude toutes les espèces ainsi que de distinguer parmi les cas de chaque espèce les possibles avec les impossibles en me fondant sur des démonstrations.'

In classifying equations Omar's first idea, not a very happy one, is to separate according to the number of terms. First come binomial equations; some roots are equal to some squares, some roots are equal to a cube. In trinomial equations his solutions of mixed quadratics are less elegant than those obtained by the application of areas, as Euclid, VI. 28 and 29. Let us compare his solution of dividing a sphere into zones of given ratio with Archimedes'. This is his fifth case of trinomial cubics. We have the equation

$$x^3 + ab^2 = cx^2.$$

We first look for the point $\{\sqrt[3]{(ab^2)}, \sqrt[3]{(a^2b)}\}$. This is Menaechmus' problem of inserting two mean proportionals between a and b

$$\frac{b}{x} = \frac{x}{y} = \frac{y}{a}.$$

We next construct a rectangular hyperbola whose asymptotes are the axes and which passes through this point. We bring it to intersect the parabola whose vertex is $(c, 0)$, whose axis is the x-axis, and whose parameter is $\sqrt[3]{(ab^2)}$

$$xy = b^{\frac{2}{3}}a^{\frac{1}{3}}; \qquad y^2 = \sqrt[3]{(ab^2)}(c-x).$$

Now let us look at some more general cubics. In order to reach an equation of the third degree from the intersections of two conics, we must either have an infinite intersection or an intersection at a known point; Omar prefers the latter. Moreover, as he is thinking in geometric terms he only wants a positive root, though negative numbers were recognized in his time.†

There are a few preliminary constructions which he explains in detail before getting down to business:

To insert two mean proportionals between two given lengths or numbers. This is the problem of Menaechmus to which we have already alluded.

To construct on a given square a rectangular parallelepiped having the volume of a given rectangular based parallelepiped. This means solving the equation $xb^2 = a^2d$, or the two equations

$$\frac{b}{a} = \frac{a}{z}; \qquad \frac{b}{z} = \frac{d}{x}.$$

This is particularly important, as his key method is frequently to construct the length a^3/b^2.

Given a rectangular parallelepiped with a square base, to construct another square-based rectangular parallelepiped of given height with the same volume as the given one. This requires solving the equation

$$cx^2 = a^2d.$$

To construct a cube of given volume.

The last two problems involve only solving certain proportions. We may summarize his scheme as follows. Suppose we wish to solve the equation

$$x^3 \pm cx^2 \pm b^2x \pm a^3 = 0.$$

First find the length a^3/b^2. Then consider the rectangular hyperbola

$$(b-y)\left(x' \pm \frac{a^3}{b^2}\right) = \pm\frac{a^3}{b}$$

and the rectangular hyperbola or circle

$$y^2 = \pm x'\left[\left(x' \pm \frac{a^3}{b^2}\right) \pm c\right].$$

These intersect where

$$\frac{b^2x'}{(x' \pm a^3/b^2)^2} = \pm\left[\left(x' \pm \frac{a^3}{b^2}\right) \pm c\right].$$

Put

$$x' \pm \frac{a^3}{b^2} = x,$$

then

$$b^2x \pm a^3 = \pm x^3 \pm cx^2.$$

† Cantor[2], p. 471.

It will be more interesting to work out three problems exactly as Omar does.

Problem 1] *A cube, some sides, and some numbers are equal to some squares,*

$$x^3 + b^2x + a^3 = cx^2.$$

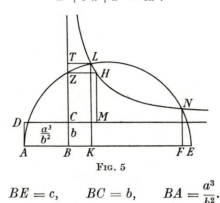

Fig. 5

Let $\qquad BE = c, \qquad BC = b, \qquad BA = \dfrac{a^3}{b^2}.$

Construct a circle on AE as diameter. Let BC meet this in Z,

$$BZ = \frac{a}{b}\sqrt{(ac)}.$$

Let $\qquad\qquad HZ.ZC = CB.BA = \dfrac{a^3}{b}.$

Construct a rectangular hyperbola with CB, CD as asymptotes passing through H. Let this meet the semicircle in L and N. One might judge from the figure that we were running into an equation of the fourth degree, but such is not the case, as one intersection of the circle and hyperbola is A, which has the coordinates $(0, 0)$, and there is one on the other branch of the curve, giving a negative root

$$LT.TC = HM.MC = BC.BA = \frac{a^3}{b}.$$

Add the rectangle CBK to both sides,

$$LK.KB = KA.BC.$$

From the circle, $LK^2 = EK.KA$,

$$\frac{BC^2}{KB^2} = \frac{LK^2}{KA^2} = \frac{EK}{KA},$$

$$BC^2.KA = KB^2.EK,$$

$$b^2KB + a^3 = KB^2(c - KB).$$

Add KB^3 to both sides,

$$KB^3 + b^2KB + a^3 = cKB^2.$$

It appears then that L is a solution, N is another. When H is way outside the circle these points are imaginary; Omar also fails to note the negative root.

Problem 2] *A cube and some squares are equal to some sides and a number,*
$$x^3 + cx^2 = b^2x + a^3.$$

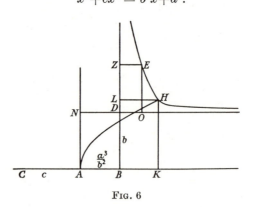

FIG. 6

Let $\qquad BD = b; \qquad CB = c; \qquad BA = \dfrac{a^3}{b^2}.$

Suppose, first, $a^3/b^2 < c$. We construct a rectangular hyperbola with vertices A and C and a second rectangular hyperbola with asymptotes DB, DN and a constant product a^3/b. The two will meet in H, where

$$HK^2 = AK\left[AK + c - \frac{a^3}{b^2}\right],$$

$$(HK - b)\left(AK - \frac{a^3}{b^2}\right) = \frac{a^3}{b},$$

$$HK\left(AK - \frac{a^3}{b^2}\right) = AKb,$$

$$\frac{HK}{AK - a^3/b^2} = \frac{AK + c - a^3/b^2}{b}.$$

But $\quad AK - \dfrac{a^3}{b^2} = BK, \qquad \dfrac{HK}{BK} = \dfrac{BK + c}{b}, \qquad BK(HK - b) = \dfrac{a^3}{b},$

$$BK^2(BK + c) = b^2 BK + a^3,$$

$$BK^3 + cBK^2 = b^2 BK + a^3.$$

When $a^3/b^2 > c$ the figure is somewhat different, but the result is the same.

Problem 3] *A cube and some numbers are equal to some squares and some sides,*

$$x^3 + a^3 = cx^2 + b^2x.$$

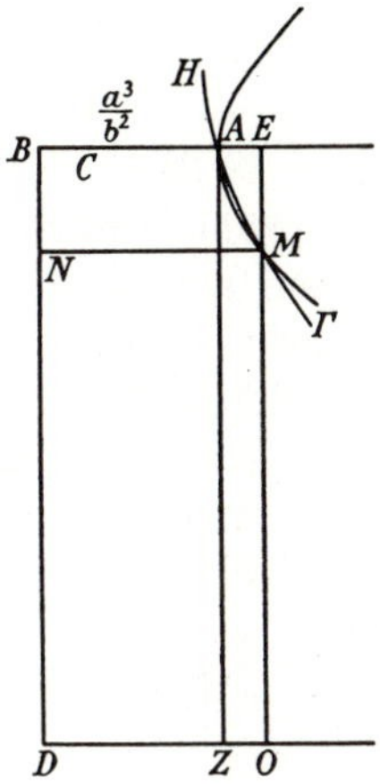

First assume (Fig. 7)

$$\frac{a^3}{b^2} > c, \qquad BC = c, \qquad BD = b, \qquad BA = \frac{a^3}{b^2}.$$

Construct a rectangular hyperbola with asymptotes BD, DZ passing through A, and a second rectangular hyperbola with vertices A and C. Let them meet in M:

Rectangle DA = Rectangle DM,

Rectangle NE = Rectangle ZE,

$$EM.MN = EO.ZO.$$

But
$$ME^2 = AE.CE; \qquad \frac{ME^2}{AE^2} = \frac{CE}{AE},$$

$$BD.AB = MN.MO = EB.ND,$$

$$\frac{BD}{ND} = \frac{EB}{AB}.$$

Subtract 1 from each side:

$$\frac{BN}{ND} = \frac{EM}{ND} = \frac{EA}{AB},$$

$$\frac{BD}{EM} = \frac{EB}{EA}, \qquad \frac{BD}{EB} = \frac{EM}{EA},$$

$$BD^2.EA = EB^2.CE,$$

$$b^2\left(EB - \frac{a^3}{b^2}\right) = EB^2(EB - c),$$

$$EB^3 + a^3 = b^2EB + cEB^2.$$

When $a^3 = b^2c$, A and C fall together. $BE = BC$, a solution. When $a^3 < b^2c$ the algebraic work is similar.

Omar passes to a lengthy discussion of equations which involve not only the three lowest powers of the unknowns but their reciprocals. He notes that here one cannot usually find a solution with the aid of conics, but in particular cases this is possible, as, for instance,

$$x^2 + \frac{a^3}{x} = cx + b^2.$$

It does not seem to me worth while to follow him here.

FIG. 7

CHAPTER III

PIETRO DEI FRANCESCHI

§ 1. The problem of perspective; Alberti

THE problem of representing three-dimensional objects accurately on a flat surface would seem naturally to be one that would attract the attention of both the mathematicians and the artists, and that at an early date. As a matter of fact both categories of persons left it severely alone until a surprisingly late time. The geometers may have thought it too simple to exercise their skill, the artists found it too mathematical to suit either their taste or their capacity. There is reason to believe that Vitruvius was familiar with some of its principles; we find him writing, 'Perspective is the method of sketching the front with the sides withdrawing into the background, the lines all meeting at the centre of the circle'.† But we must wait for the Italian Renaissance, which produced a number of extraordinary men interested alike in the arts and the sciences, for a detailed study of the really simple principles involved.

The credit for discovering the first principles of perspective is usually given to Bruneleschi. We read in Vasari:

'Filippo Bruneleschi gave considerable attention to the study of perspective, the rules of which were then imperfectly understood, and often falsely interpreted; and in this he expended much time until at length he discovered a perfectly correct method, that of taking the ground plane and sections by means of intersecting lines, a truly ingenious thing and of great utility in design. . . . This work having been highly commended by the artists and all who were capable of judging matters of the kind gave Filippo so much encouragement that no long time elapsed before he commenced another and made a view of the Palace, the Piazza, the Loggia del Signori with the roof of the Pisani and all the buildings erected around the Square, works by which the attention of artists was so effectively aroused that they afterwards devoted themselves to the study of perspective with great zeal. To Masaccio in particular, who was his friend, Filippo taught this art.'‡

The statement here of what Bruneleschi actually discovered is obscure enough. For some time painters had been feeling for a theory of perspective and their results were by no means too grotesque in practice, even if they had no theory. A long discussion of the point will be found in Bunim (q.v.). We may well accept the statement that after Bruneleschi's time the study of the subject became popular, for 'In 1435, a few years after the fresco was painted, the first treatise in which the method of focussed perspective, designed for the use of painters, was written by Alberti.'§

† Vitruvius (q.v.), p. 14.   ‡ Vasari (q.v.), vol. i, p. 249.   § Bunim (q.v.), p. 187.

The work of Alberti was certainly intended for followers of the painter's art, the intention appears on every page. For example:

'And indeed I shall think I have done enough if Painters, when they read me, can gain some information on this difficult subject which has not, as I know of, been discussed by any author. . . . And because it would be tedious as well as extremely difficult and obscure in this method of intersection of Triangles and Pyramids, to handle everything in a mathematical way, we shall pursue our discourse according to the custom of painters.'†

The work begins with a general description, with excellent figures, of the principles of foreshortening. He shows clearly how nearer objects appear larger and, in general, discusses what he calls pyramids of rays. Alberti wrote for members of his own craft; let me claim the same privilege, and give a few definitions according to the custom of the mathematicians.

A picture is supposed to be a drawing to scale of what exists in a supposed 'picture plane' standing upright on the 'ground plane' between the artist and the objects which he represents. This plane is pierced by rays of light going straight from the objects to the artist's eye, so that we have in the picture plane a central projection of the objects in question. The essential constants in the construction are the height of the artist's eye and the distance from that eye to the picture plane. The position of the eye in space is called the 'station point'. The foot of the perpendicular from the station point on the picture plane is called the 'centre of vision'. A vertical plane through these two points is called the 'central plane'; it intersects the picture plane in the 'prime vertical', the vertical line through the centre of vision. The intersection of the picture and ground planes is called the 'ground line'. A horizontal line in the picture plane through the centre of vision is called the 'vanishing line' or 'horizon line'. The two points on this line the same distance from the centre of vision as the station point are called the 'distance points'. A straight line in space will always appear as a straight line in the picture; parallel horizontal lines, wherever they may be, will appear as lines meeting on the horizon line; lines perpendicular to the picture plane will appear as lines through the centre of vision.

Alberti's main problem is to represent in the picture plane a set of squares in the ground plane whose sides are parallel to, or perpendicular to, the ground line. We first mark off a set of equal distances on the ground line and connect them with the centre of vision. These will give one set of sides. More difficult is it to draw lines parallel to the ground line which shall represent equally spaced lines parallel thereto. Alberti tells us that troubled his predecessors not a little. One suggestion was

† Alberti (q.v.), pp. 1 and 7.

to space them according to a descending geometrical progression. He comments: 'This method may be practised by some, but if they imagine it a good one, I am of the opinion they are not a little mistaken. . . . The matter standing as I have shown, I have myself invented the following method which I have found, by experience to be a very good one.'†

I exhibit this in Fig. 8, which is taken photostatically from Alberti. I will not give his proof, but save a good deal of labour by introducing

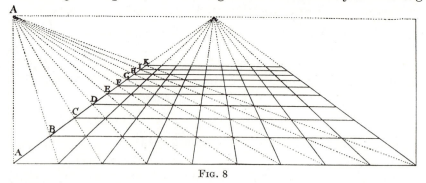

FIG. 8

algebra. Let the height of the station point be h, the distance to the picture plane a. If a line in the ground plane be parallel to the ground line at a distance x therefrom, it will appear in the picture plane as a line parallel to the ground line whose height above the latter is

$$y = \frac{hx}{x+a}.$$

We swing the central plane about the prime vertical till it lies in the picture plane. A superior is thus the distance point. If we take AA and the ground line as rectangular Cartesian axes, the line $ABCD...$ will have the equation

$$\eta = \frac{h}{a}\xi.$$

If we connect $(x, 0)$ with the point A superior whose coordinates are $(0, h)$ the connecting line will meet $ABCD...$ in a point for which

$$\eta = \frac{hx}{x+a}.$$

This proves the correctness of the construction.

§ 2. Pietro's first method

It seems to me perfectly certain that Alberti's knowledge of perspective did not cease with the representation of a chequerboard of squares, but that is all that is given in this work. The man who first set forth the principles in fairly complete form was Pietro dei Franceschi, or

† Alberti (q.v.), p. 8.

Pierro della Francesca as he is frequently called. It is hard to say just how original he is. There is some discussion of the point by Winterberg in the introduction to Franceschi (q.v.). This translator's thesis is that Franceschi merely set forth principles already familiar to artists, especially architects. I am not entirely convinced of the correctness of this view. Franceschi's work has been placed somewhere between 1470 and 1490, Alberti's dated from 1435. It is hard to believe that this writer's technique had become current practice in so short a time. Pietro was not a great mathematician, and he certainly did not write for mathematicians, or in the style they affect. But he had read the works of great geometers of the past. He explained two very different methods for constructing perspective figures, one following Alberti's model in essence, the other entirely different. He may have invented it, he may have taken it from some unnamed source, but he was the first writer to set forth in great detail the methods for meeting all sorts of problems in perspective which may arrive in practice. Pittarelli speaks of 'L'opera sua come scrittore di un trattato completo di prospettiva, il primo che vedesse lume in Italia e nel mondo.'† I have personally found him hard to follow, even in Winterberg's translation, he uses many non-mathematical terms. I therefore reproduce some of his excellent drawings and try to reproduce his thought from them.

Pietro begins with a number of simple geometrical principles and drawings which illustrate the principles of foreshortening. He then goes on to representing in the picture plane certain points and figures in the ground plane. He imagines that this ground plane, or so much of it as lies beyond the picture plane, has been swung around the ground line as an axis until it lies on the picture plane, just below that part in which he will draw his picture. We have a correct drawing in the rotated ground plane, and just above it the representation in the picture plane. Let us see what is going on. If we follow Alberti's method we can find the point in the picture plane corresponding to a given point in the ground plane as follows. In the picture plane we mark the centre of vision, distance point, ground line, and prime vertical. Through their intersection we may draw in the ground plane a line making an angle of 45° with the ground line. We wish to represent this in the picture plane; we can do so if we can represent one of its points. Choose a point and draw through it a perpendicular meeting the picture plane in a point, say s. Connect this with the distance point and centre of vision. Find where the former line meets the prime vertical and through there draw a parallel to the ground line. This will meet the line from the centre of vision to s in the point representing the point of the 45° line. But Pietro does not like to do it in this way, he does not like to represent

† Pittarelli (q.v.), vol. xii, p. 254.

the distance point. If we have marked the distance point we can, as we have seen, draw the line in the picture plane which represents any particular line in the ground plane parallel to the ground line. Conversely, if we know what line in the picture plane represents any particular line in the ground plane parallel to the ground line, and the

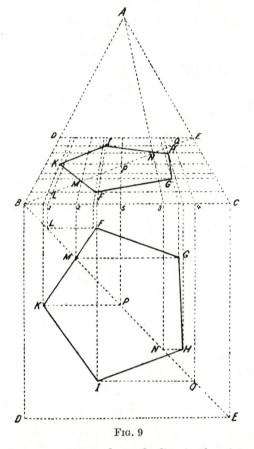

FIG. 9

centre of vision, we can at once draw the line in the picture plane which represents the 45° line in the ground plane. Suppose, then, that we have these data, and a point marked in the ground plane. We draw a vertical line through it to meet the ground line, and connect the intersection with the centre of vision. This will represent the vertical line through the given point. Again, through the given point draw a line parallel to the ground line to meet the 45° line, and find by the method just described the representation in the picture plane. This will enable us to draw in the picture plane that line parallel to the ground line which represents the corresponding line in the ground plane. We have, thus,

the representation of two lines through the given point; their intersection will represent that point. In Fig. 9 we see how Pietro represents
a regular pentagon in the ground plane by this device.

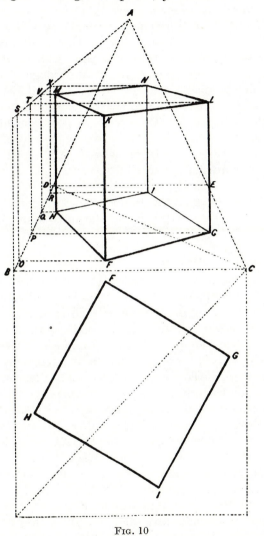

Fig. 10

Unfortunately the objects which we wish to represent do not always
lie in the ground plane. Pietro next undertakes to show how to represent
points lying above the ground plane. It is inconceivable that Alberti
was unable to do this, but he does not show us in the work in question.
The difficulty arises from the fact that distances in the picture plane
above the ground line come from distances above the ground plane and

distances behind the picture plane. Here is Pietro's procedure. I wish to represent a point which is a height h above the ground plane, and whose shadow thereon is known. We draw in the picture plane a line parallel to the ground line at a height h above it, pick a point z thereon (not marked in Fig. 10), and drop a perpendicular zB to the ground line. We connect B with A, the centre of vision. How shall we represent the point which is at the height h above that point which is represented by the point F in the picture plane? We draw through F a line parallel to the ground line to meet AB in O. Erect a vertical line at O to meet Az in s. Through s draw a line parallel to the ground line to meet the vertical line through F in K. Then K is the point sought. We see, in fact, that $sO/h = AO/AB$, which ratio is independent of the position of B on the ground line, and if we place B at the foot of the prime vertical and consider what happens in the central plane, we see that we have the right value. In Fig. 10 we have Pietro's representation of a cube of height h standing on the ground plane. He follows with eleven other figures showing the method of representing columns, prisms, arches, and other vertical objects.

§ 3. Pietro's second method

At this point Pietro turns completely about and shows a second method of finding the picture. He says that this really amounts to the same thing, which of course it must if each is merely a method for finding the intersections of the picture plane with lines and planes radiating from the station point. It is my impression that he feels the first procedure would clutter up the picture with too many construction lines, while the work of the second method can be done on other sheets. The important point which I have not been able to settle is what is the source of this new method. It certainly does not stem from Alberti's work. Winterberg, in Franceschi (q.v.), does not seem to appreciate how complete is the change. It would be pleasant to find herein a really original contribution to mathematical science, but was Pietro a sufficiently capable mathematician to accomplish this? Let us give him all possible credit anyway.

I must confess to finding this part of Franceschi[1] even harder to understand than what precedes; there are, of course, no real mathematical proofs. An additional difficulty arises from the introduction of new mathematical instruments, strips of wood or paper, needles, threads, strings. These, of course, are only brought in for geometrical purposes. Fortunately his drawings are very clear. I shall therefore, as before, try to reconstruct from the figure the geometrical statement of what is accomplished.

Pietro lived long before the time of Fermat and Descartes, and certainly had no general idea of rectangular coordinates; nevertheless his second method consists in calculating from the original data what will be the coordinates of the representing points in the picture plane. We take the ground line as axis of abscissae, the prime vertical as axis of ordinates. Abscissae are marked on the ground line, ordinates on two vertical lines out of the way of the picture. With a draughtsman's triangle we may erect a perpendicular to the ground line at any chosen point, and with a ruler connect the two corresponding points on the vertical lines. Where they cross is the point whose coordinates are given. In practice the two coordinates are handled separately.

The abscissa of the representative of a given point is that of the point where the ground line meets a vertical plane through the given point and the station point. This plane contains the orthogonal projections on the ground plane of the given point and the station point. If, then, we connect these two projections, the connecting line will meet the ground in the point whose abscissa will give us what we want.

The determination of ordinates is slightly harder; Pietro employs different techniques in different problems. Let us first note that all points in space on a line parallel to the ground line will be represented by points having equal ordinates; that is to say, points of another line parallel to the ground line. We may find the ordinate of a point in the ground plane in this way. Draw a line, which I will call the line l, perpendicular to the ground line at a distance from the ground point equal to the height of the station point. To find the ordinate of the representative of a point P in the ground plane, we draw PQ perpendicular to the line l, connect Q with O, the projection of the station point on this plane. The connecting line will meet the ground line in a point whose distance from l is equal to the ordinate sought; the proof comes at once by similar triangles. We find the ordinate of a point representing a point not in the ground plane by replacing it by the point where the ground plane meets the line from the station point to the given point. In Figs. 11 and 12 we have Franceschi's picture of a regular octagon lying in the ground plane, one side being on the line l.

A more difficult problem is that of representing a cube in general position. This we have in Fig. 13. Let us begin with the abscissae. The vertices of the cube shall be $ABCDFGHI$. I shall assume that the face $AFID$, which I shall call the base, makes an angle θ with the ground plane, and that the line of intersection is vertical in this picture. If we rotate the base plane about the line of inter-section until it lies on the ground plane, we have the figure in the south-west corner of Fig. 13. The projections on the ground plane of the four edges AB, FG, DC, IH when in original position will be four equal horizontal segments on four horizontal lines through the

rotated *AFDI* in the south-west corner. If we project the cube when
rotated to the south-west position orthogonally upon the central plane
we get a figure which when rotated down appears in the north-west

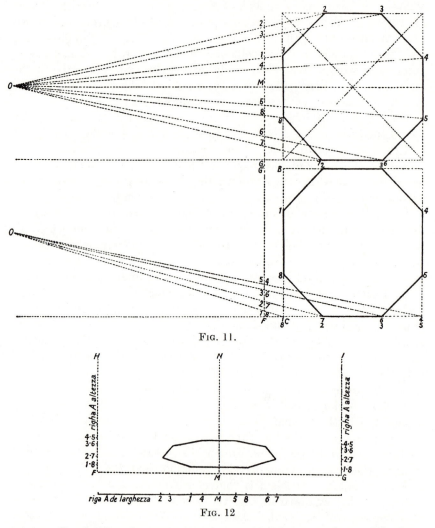

Fig. 11.

Fig. 12

corner. In the north-east corner we take *A* at random. We draw
through it a line segment *ADIF* equal to the segment of that name in
the north-west, making an angle θ with the vertical, and *AB* perpendicular to it equal to *AB* in the north-west. In other words, we reproduce the north-west figure in the north-east so that *AF* makes an angle θ
with the vertical. Through the eight points in the north-east we draw
vertical lines to meet the corresponding horizontal lines in the south-east

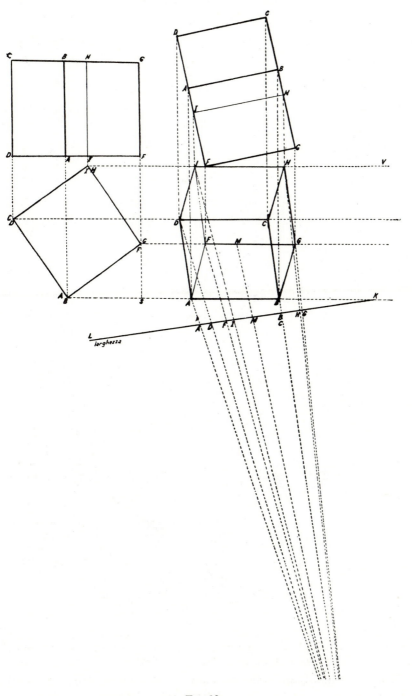

FIG. 13

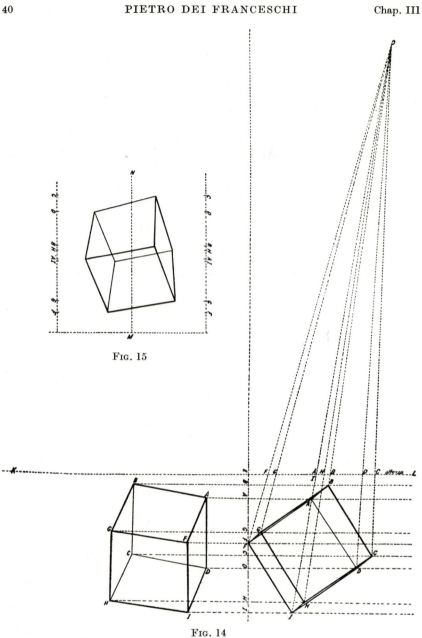

Fig. 15

Fig. 14

in the points which give the orthogonal projection of the cube on the ground plane. *LR* is the ground line.

The determination of the ordinates is less satisfactory. Frankly I cannot make any sense out of Pietro's own statement. The plan here is to project the cube horizontally on the central plane. In Figs. 14 and 15 KC is the prime vertical, O the station point, the plane of the paper is the central plane. To project a point on the central plane we might first project on the ground plane, then rotate that about the intersection till it stood on the central plane, then slide the point downwards the proper distance. That is what Pietro has done. The figure in the south-west of Fig. 14 is that in the south-east of Fig. 13 slewed around. The eight points are then slid horizontally to the position in the south-east. I do not understand his rule for determining how far to slide them, and I doubt whether this figure in the south-east really is the orthogonal projection of the cube on the central plane. The faces $ABPG$ and $CDHI$ are flattened nearly to line segments. I doubt very much whether this would happen, judging from the other figures. But I think the author deserves real credit for what he has accomplished, no small advance theoretically at least, on the work of his predecessors.

§ 4. *De corporibus regularibus*

I may not leave Pietro without a short reference to his other mathematical work, Franceschi[2], *De corporibus regularibus*. As the title indicates, he is occupied with the five regular solids. He begins with a number of applications of the Pythagorean theorem. When he comes to the regular pentagon he gives the ratio of the square of the diameter of the circumscribed circle to the square of one side in the correct if somewhat unusual form $16/(10-\sqrt{20})$. He takes $\pi = 3\frac{1}{7}$. When it comes to inscribing regular solids he confesses his indebtedness to Euclid XIII, though what he has written is much easier reading than the work of the master. He ends up with a decidedly interesting problem. This is to find the volume of the solid contained by two equal cylinders of revolution, whose axes cut at right angles. This problem was set by Archimedes and solved by him on p. 48 of Archimedes[3]. But this part of Archimedes' work was long lost and first recovered in 1907. This seems to me Pietro's best piece of mathematical work. I incline to think that he invented and solved the problem first himself.

There has been rather an absurd amount of controversy over the question of whether Luca Paccioli's *De divina proportione* is not a shameless copy of the work, Franceschi[2]. Vasari considered Paccioli as a perfectly unblushing thief of other men's work. Waters (q.v.), in his book about Pietro gives reasons to doubt this. But others have entered the fray, Chasles, Cantor, Steigmüller, and Pittarelli. Loria and Volterra in their introduction to Pietro Franceschi[2], lay stress on the fact that the same numbers are used in giving examples of geometrical facts, though there was no inherent reason for the choice. I have not been able to find the examples they suggest. A charitable view is expressed

by Mancini in his preface to Pietro[2]. He gives reasons for believing that
Paccioli copied, but excuses him on the interesting ground that in his
time this was not considered a very blameworthy thing to do. I have
not gone as deeply into the matter as he has, but a hasty study leaves
me rather sceptical. Pietro seems very little interested in the *Divina
proportio* or 'golden section' as we call it, the division into extreme and
mean ratio. He mentions it casually, as he must, in connexion with the
pentagon. Luca turns it about in every way, bringing in again and
again the mystic numbers $\sqrt{125}-5$, $15-\sqrt{125}$ which I do not find
explicitly in Pietro. He pays a good deal of attention to the inscription
of the regular bodies in one another, which does not appear in Pietro,
but he gives nothing as interesting as Pietro's problem of the volume
between two cylinders which I mentioned above. The question does
not seem to me vital, the whole subject was started long before by
Euclid anyway.

CHAPTER IV

LEONARDO DA VINCI

§ 1. Background

OF all the sons of men of whom we have any record, none had a greater intellectual appetite, or shall we say, a more omnivorous curiosity, than the great Florentine whose name appears at the head of the chapter. He is the closest approach we have to that unattainable ideal, the universal man. The Italian Renaissance produced a number of men who stood out both in science and art, but Leonardo was supreme. We think of him first as a really great painter, and that opinion will probably hold good until the end of time, but in that connexion he was deeply interested in anatomy, both human and animal, and is said to have dissected a large number of human bodies. It has been suggested that the famous intriguing smile of La Gioconda could only have been painted by one well acquainted with the muscles of the mouth; I cannot see exactly how it would be possible to prove such a statement. His treatise *Della pittura* clearly shows his interest in anatomy, but also his absorption in mechanics. Mechanics led to the construction of war machines, a highly prized accomplishment in that warlike time, but it also led to the study of the accomplishment of human flight. Leonardo was convinced, perhaps wrongly, that the human frame had sufficient muscles for flight, it was only a question of attaching them to the proper mechanism. He made careful observations of the flight of birds, and gave careful consideration to the essential question of the relation between the centres of gravity and wing pressure. He could not foresee the invention of the internal-combustion engine which alone has made flight possible for us. The mechanics of rigid bodies borders on that of fluids; he made both theoretical studies and practical undertakings in the science of hydraulics. Physics is next to chemistry, he was much preoccupied with the mixture of paints, although here, as in other places, his ventures were not uniformly successful; his great unfinished painting, the battle of Anghiari, was completely spoiled when wrongly mixed paint ran, under the influence of heat. A more elaborate list of his scientific activities will be found in Libri (q.v.) or Marcolongo[1], the great collections of his scattered notes and designs, as Leonardo[1] and Leonardo[4], contain an unbelievable wealth of ideas and suggestions dealing with all manner of subjects.

Our present interest is in Leonardo as a mathematician, although this is not one of the fields where he showed marked superiority. A careful and detailed study will be found in Marcolongo[1] and Marcolongo[2], especially the former. If I were to venture on a criticism of the work of this distinguished mathematician it would be to say that he was so much impressed by the universal genius of his extraordinary fellow countryman that he was over ready to excuse errors and omissions which seem to me important; but when it comes to lavishing praise, he is far behind Libri. Be that as it may, short of giving many years to the study of Leonardo's mathematics, it would be hard to add anything substantial to what Marcolongo has written. I will make this exception: he pays no attention to Leonardo's inscription of regular polygons, which seems to me interesting, and to which Cantor gives considerable attention.

What chance did Leonardo have to learn mathematics? There is a detailed documentary study of this in Marcolongo[1].† His early education was certainly not carried far, but his insatiable appetite drove him to seek information whenever and wherever it could be found. During his apprenticeship in the studio of Verocchio he met men of outstanding ability, and throughout his life he had the most stimulating intellectual contacts. In particular he was intimate with Luca Paccioli, whom I mentioned on p. 42, and later he was in contact with Nicolo Cusano. I should also mention Leonbattista Alberti, and Pietro dei Franceschi.

How about his mathematical reading? There were parts of Euclid that he knew well, especially the Pythagorean theorem, and the construction of a mean proportional. With other parts of Euclid he does not seem to have been familiar, or they did not interest him. He makes no mention of the gnomon or the Application of Areas, and I doubt whether he really appreciated the method of exhaustion, although that is much in evidence in the writing of another Greek mathematician whom he greatly admires, Archimedes. He certainly was familiar with the encyclopaedic work of Giorgio Valla (q.v.). This writer gives a long study of the problem which intrigued Leonardo, the insertion of two mean proportionals between two given lengths, or numbers. He gives in turn a long list of attempts at solution by various Greek geometers. This list first appeared in Eutocius' commentary on Archimedes' work on the sphere and cylinder.

The mathematical work of Leonardo which I shall discuss in detail falls under five heads:

    I. The areas of lunes.

    II. Solids of equal volume.

    III. Reflection in a sphere.

    IV. Inscription of regular polygons.

    V. Centres of gravity.

† pp. 31–47.

§ 2. Areas of lunes

Hippocrates of Chios, a mathematician of the fourth century, the first Greek to prepare a book of elements, made the surprising discovery that there are certain plane figures, bounded by circular arcs, whose areas could be calculated, that is to say, we could construct geometrically squares having the same areas. This discovery stimulated a great deal of research, quite out of proportion in fact to its own intrinsic importance, but the problem of squaring the circle had, from earliest times, been so alluring that anything that looked in that direction was bound to arouse interest. If we can find exactly the area of a figure bounded by two circular arcs of different radius, why not that of the simpler surface bounded by a single circle ? Here are Hippocrates' two theorems:

I) If a semicircle be described on the hypotenuse of an isosceles right triangle as diameter so that it passes through the vertex of the right angle, and another on one of the legs, also passing through that vertex, the area of the 'lune' between the two is half the area of the triangle.

II) If a semicircle be constructed on a diagonal of a regular hexagon which connects a pair of opposite vertices, and additional semicircles on the three sides of the original hexagon which are chords of the semicircle, the sum of the areas of the three lunes between the large semicircle and the three small ones, plus that of one small semicircle, is half of the area of the original hexagon. When it comes to a proof we have merely to note that the areas of semicircles are to each other as the squares of their diameters, so that in the first case the area of a small semicircle is half that of the large one ; we then take from each the common segment. In the second case the area of a small semicircle is one quarter that of the large one ; we then subtract the areas of the three small circular segments.†

It is perfectly evident that each of these theorems can be easily generalized. Curiously enough, progress in this direction came slowly. The first important step was to consider a right triangle which was not isosceles, and to describe semicircles on each of the legs and on the hypotenuse. Then as the sum of the squares on the legs is equal to the square on the hypotenuse, so the sum of the areas of the two small semicircles is equal to the area of the large one. We then subtract the sum of the two small segments and reach the pretty theorem that the sum of the areas of the two lunes is equal to that of the right triangle. Now this was first discovered about the year 1000 by Hasan ibn al Haitam, as we learn from Suter (q.v.). Exactly what drew Leonardo's attention to it we do not know ; it is hard to imagine that he ever saw

† Cf. Marcolongo, p. 52, and Heath (q.v.), vol. i, pp. 183 ff.

the text of the Arabian author which was first translated into a European tongue in 1899. I reproduce Leonardo's bad drawing in Fig. 16. Here is the text:

'Qui sempre li 2 semicirculi *a*, *b* insieme giunti sono eguali al terzo dove è fatto l'ortogonio. E se a cose equali si leva la parte equale, il rimanente

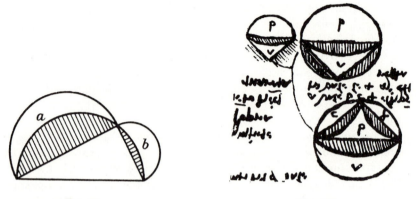

Fig. 16   Fig. 17

saranno equali. Se dunque, che tolte il depennatto (che e doppio) allo *a* e tolto al *b* restano le lunelle, e di poi tolto il depennato al semicirculo maggiore *n* che vale ai due predetti, seguita che un ortogonio resta equale alle due lunelle.'†

It is noteworthy here that we learn next to nothing about the area of one lune, it is the sum of the areas that appears.

It is perfectly incredible how Leonardo threw himself upon this not too important theorem, deducing therefrom numberless particular instances of rectifiable figures bounded by circular arcs, or dissimilar figures of the same area. On one page I have counted 176 individual figures, a photostatic reproduction would be too confusing. His various procedures are somewhat codified in Marcolongo[1]. If similar figures are constructed on the legs and hypotenuse of a right triangle, the sum of the areas on the legs is equal to the area on the hypotenuse. An example of this appears in Fig. 17. It is to be noted here that the segments have angles of 45°. If we construct a semicircle outside on one of the legs we get a lune with an angle of 45°. If we rotate this lune about one corner through an angle of 45°, we get an eight-petalled flower of known area inside a circle.

One simple method of finding a figure of known area is to take a polygon with two equal sides, gouge out a figure of any shape inside one of these sides, and add an equal figure outside the other. I give two

† Marcolongo, p. 54.

examples in Fig. 18. In Fig. 19 we have something more elaborate; it is taken from Leonardo[1], 98 v. Let the radius of the circle inside the square be 1, the radius of a small semicircle is $\sqrt{2}/2$. The area of each small segment is one-half that of a large segment. The area of the out-

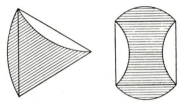

FIG. 18

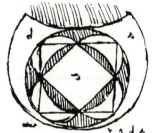

FIG. 19

FIG. 20

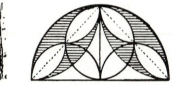

FIG. 21

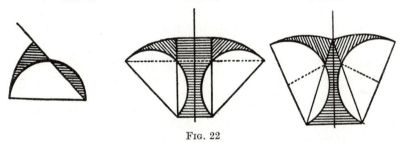

FIG. 22

side circle is 2π. The area of the four large segments is $\pi-2$. Hence the shaded area is $2\pi-4$ and the non-shaded is 4.

The area of each petal in Fig. 20 will be $\frac{1}{4}\pi-\frac{1}{2}$; the shaded area will then be $\pi-2$ and the non-shaded 2.

Leonardo was rather fond of figures inscribed in semicircles. We have a good example in Fig. 21, also from Marcolongo[1]. The area of a large petal which is four times that of a small one is $\frac{1}{2}\pi-1$. The area between the two large petals and the small ones is the same, so that the area of the rest of the big semicircle is 1.

Leonardo makes much use of the fact that the area of one-eighth of a circle of radius 2 is that of one-half of a circle of radius 1. In Fig. 22 we see that by taking away from both the common area we have a horn-shaped figure of the same area as a segment of a circle. Here are two examples from Marcolongo[2],† though the drawings are based on Leonardo's own. In one case the shaded part has the area of a rectangle, in the other case of a triangle. I note that in Marcolongo[1],‡ this is handled wrongly. One could go on almost indefinitely giving such examples; I can but wonder that such a great man paid so much attention to what was, after all, a pretty unimportant matter, his ingenuity in thinking up examples is, of course, extraordinary.

§ 3. The transformation of solids

Leonardo was not a professional mathematician; he did not choose his problems from among those most likely to advance mathematical science, otherwise he would not have spent so much time and strength on figures bounded by circular arcs. Another rather curious fancy was the transformation of solids into others of equal volume. The Greeks had, so to speak, completely solved the analogous problem in two dimensions. They were even able to construct a polygon, similar to a given polygon, and having the area of another given polygon. Transformation of solids had already been studied by Nicolo Cusano. Leonardo saw therein more or less of a practical problem, which to his mind naturally made it more interesting. We find him writing: 'Geometria che s'astende nelle trasmutazioni dei corpi metallici che son di materia atta a astendersi e racortare secondo la necessità de loro speculanti'.§ His whole point of view is essentially practical. He prefers constructions which can be found experimentally to those which can be demonstrated rigorously. It appears certain also that he wrote a treatise on this subject which has been lost, for in Leonardo[2] we find references to such numbers as 'la 5ᵃ del 2°, la 6ᵃ del p°'.

The solids with which he is principally occupied are rectangular parallelepipeds with square bases. When such a solid has a height greater than the length of the base he calls it a *cylindro*; when the height is less it is a *tavola*. It is noticeable that Leonardo[2] is, so to speak, written backwards; not only that, like most of Leonardo's work, it was mirror-written, but that the more elementary parts on which the others depend come at the end. Fortunately in Marcolongo[1], p. 312, we have the essential material arranged in proper order. Let us begin with Leonardo[2], 38 v., where he finds a square equivalent to a given rectangle. This, as Leonardo himself acknowledges, is Euclid, II. 14; he even copies Euclid's figure. On p. 59 he mentions the reverse problem, that of finding a rectangle of given length equivalent to a given square, also that of finding a rectangle equivalent to the sum of two rectangles. There is nothing of interest here.

† p. 163.   ‡ p. 62.   § Leonardo[2], 40 v., p. 70.

We find something more worth while in 34 v., p. 58, to construct a cube equivalent to a given cylinder. This involves solving the equation

$$x^3 = a^2 b,$$

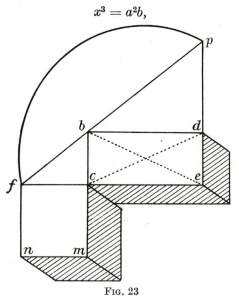

FIG. 23

which is the same as the introduction of two mean proportionals

$$\frac{a}{x} = \frac{x}{y} = \frac{y}{b},$$

the most famous problem in Greek mathematics, and the source of our knowledge of the conic sections. Leonardo was certainly familiar with Valla (q.v.), who follows at length Eutocius' commentary on Archimedes, where many solutions are given.† Leonardo chooses that of Hero. In Fig. 23 we have his rather crude drawing. Let *bcde* be a face of the cylinder. We take as a centre the intersection of the diagonals, and draw two circles; one, not shown, through *b, c, d, e,* the other meeting *ce* and *de* in two such points *f, p* that *fp* passes through *b*. We know to-day that this cannot be effected with the aid of ruler and compass alone. As for a proof we note that

$$fc:cb = fe:ep = bd:dp.$$

Now as *f* and *p* are at the same distance from the centre of the circle, they have the same power with regard to it, so that

$$fc.fe = ep.dp$$
$$fe:ep = dp:fc$$
$$cb:fc = fc:dp = dp:bd.$$

† Archimedes², vol. iii, pp. 71, 72.

On the preceding page we have the problem of solving the equation

$$m^3 = bx^2,$$

$$fc = m, \qquad cd = b.$$

Erect (Fig. 24) perpendiculars to cd at c and d, and n, the middle point.
On this last perpendicular nh we find (he does not tell us how) such a

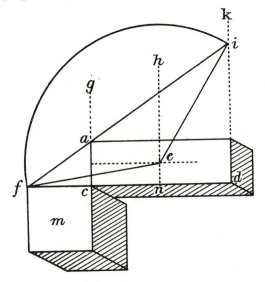

Fig. 24

point e that if a circle be drawn with e as centre passing through f it
will meet db in k and fk will meet the perpendicular at c in A, so that
$cA = 2ne$. We now have the same arrangement as in the last figure,
and may proceed as before. We find a different proof in Marcolongo[1].†
We have finally the problem of solving

$$m^3 = a^2x,$$

and here again he works around to the same figure.

One wonders, naturally, why Leonardo follows such a clumsy technique. The obvious way would have been to take the last two problems
in reverse order. He knew perfectly well how to find y so that

$$m^2 = ay,$$

$$m^3 = amy = a^2x,$$

$$x:y = m:a.$$

As a matter of fact, he does give such a solution in another place.‡ As
for the problem
$$m^3 = bx^2$$

† p. 314.   ‡ Marcolongo[1], p. 314.

we find y so that
$$m^2 = by,$$
$$m^3 = bmy = bx^2,$$
$$my = x^2.$$

This is a problem already solved. It is clear, I think, that Leonardo's object is to find a uniform method which a clever workman can follow

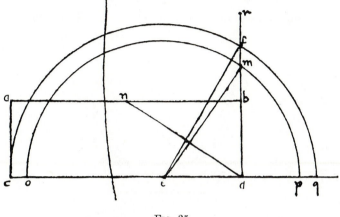

Fig. 25

experimentally rather than an exact one. As a matter of historical fact all of these problems save the first were solved in Euclid XI.

Leonardo gave applications of his general methods to certain specific problems. In Leonardo[2][†] we have the good problem of constructing a rectangular parallelepiped, equivalent to a given cube, whose measurements are in given ratios. This amounts to

$$x/b = y/c = z/d; \qquad xyz = a^3.$$

We find in turn
$$bcd = bs^2 = r^3,$$
$$a/r = x/b = y/c = z/d.$$

This is a correct solution, but Leonardo was by no means above making mistakes. In Leonardo[2][‡] we have another attempt to solve for x the equation $yz^2 = x^3$. I change the lettering from the equation on p. 50 to avoid confusion with the rather obscure Fig. 25.

Let
$$cd = y ; \qquad db = dq = z.$$

Let e be the middle point of cq:
$$df = \sqrt{(yz)}.$$

† p. 24.   ‡ F. 14 v., p. 19.

Let n be the middle point of ab:

$$ed = \frac{y+z}{2} - z = \frac{y-z}{2},$$

$$nd = \sqrt{\left(\frac{y^2}{4} + z^2\right)} = ep \quad \text{by definition of } p,$$

$$md = \sqrt{(ep^2 - ed^2)} = \tfrac{1}{2}\sqrt{(2yz + 3z^2)}.$$

For some strange reason he says that this is the value of x.

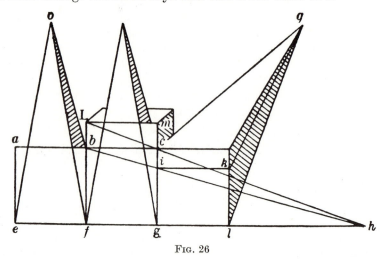

Fig. 26

Leonardo is interested in pyramids as well as in parallelepipeds. He realizes that the volume of a pyramid is one-third the product of the base and the altitude. His problem† is to find a pyramid of given height whose volume is that of a given cube. The figure Leonardo[2], p. 37, is too bad so I follow that on p. 325 of Marcolongo[1] which, of course, is based on the original. The squares $abef$, $bcfg$ are typical faces of the cube. Construct a pyramid of the given height, which we assume greater than three times the length of an edge of the cube, on a face of the latter. Construct a rectangular parallelepiped on the same base with one-third the height of this pyramid. This shall be $Lmfg$. We wish to find a rectangular based pyramid of height $3Lf$ and volume $(bf)^3$.

Draw Lc to meet fg in h. Let bh meet cg in i:

$$Lf/bf = cg/ig = bf/ig,$$

$$Lf.ig = (bf)^2; \qquad Lf.ig.bf = (bf)^3.$$

I mention in conclusion that Leonardo sometimes really went off the deep end, at least if he was trying to do exact mathematics. I take as

† Leonardo[2], F. 24 r., p. 38.

an example Leonardo[2], p. 12: 'Di un cubo sia fatto un corpo di 12 eguali base pentagonali.' The problem of finding a regular dodecahedron equivalent to a given cube was surely above his depth.

§ 4. The spherical mirror

In his long treatise on painting Leonardo showed great interest, not only in shadows, but in general questions of illumination. I note in passing that it was Leonardo who made the discovery that the familiar phenomenon of 'The old Moon sitting in the new Moon's lap' was caused by the illumination of the Moon by Earthshine. It was consequently natural that such a man should become interested in the problem of the spherical mirror. What is that? It is the problem of finding the point of impact on a spherical mirror of a ray issuing from a given point and arriving at a given point, after reflection. But right at the outset we are faced again with the same problem of originality as before, for this problem also had been stated, if not completely solved, by Hasan ibn al Haitam. It is a suspicious circumstance that Leonardo should have taken two problems from this author; in the first case, as I said on p. 46, it seems to me that we can excuse the Italian from any charge of copying; not so here. The Arabian author's treatise had been translated into Italian in the fourteenth century,[†] and may have been available to Leonardo in Pavia or in the library of San Marco in Florence. Moreover, a treatise on optics based on his work by Vitellio was available, and had been consulted by Paccioli.[‡] Finally Huygens, to whom we must turn for a good solution, in a letter written in 1669 speaks of Problema Alhesini.[§] It seems to me therefore that although this was a problem that Leonardo might easily have thought of for himself, he really took it from the Arabic source.

How about the solution? Leonardo himself tried hard and failed. Let us see what is involved. It is a two-dimensional, not a three-dimensional problem, for the point sought must obviously be in the plane through the centre of the mirror, the source of light, and the eye; it is a question of a circular, not a spherical mirror, and looks very easy at first sight. But when we look further into the matter difficulties appear. It is the problem of finding where the circle will be touched externally by an ellipse whose two foci are the source and the eye, and we find that analytically there will be four solutions. These were worked out by Huygens and others. I will give the neatest form due to Catalan.[||]

Let us take the lines from the centre to the source and the eye as x- and y-axes for a set of oblique Cartesian coordinates. The inverse of

† Narducci (q.v.), pp. 1 ff.   ‡ Marcolongo[1], pp. 72 ff.
§ Huygens (q.v.), vol. vi, p. 462. || Oznam (q.v.), p. 486.

the source shall be $(x_1, 0)$ that of the eye $(0, y_1)$, the point sought (x, y). We have two similar triangles, the one with vertices (x, y), $(x, 0)$, $(x_1, 0)$, the other with vertices (x, y), $(0, y)$, $(0, y_1)$:

$$\frac{y}{x-x_1} = \frac{x}{y-y_1},$$
$$x^2 - y^2 = xx_1 - yy_1.$$

We have a hyperbola, which is rectangular, through the centre and the inverses of the eye and the source.

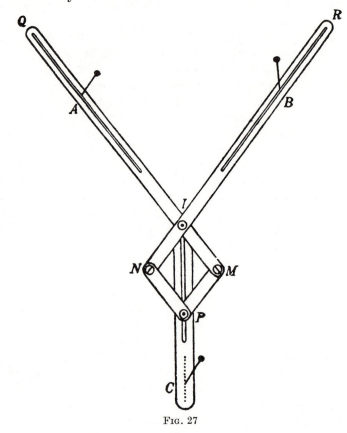

FIG. 27

Leonardo himself made various attempts at a mathematical solution and failed, but he was by no means a man to be satisfied with failure. He proceeded to devise a simple mechanical scheme for accomplishing his end. This we see in Fig. 27. The picture in Leonardo[1]† is unsatisfactory, so I copy that in Marcolongo[1],‡ which represents a machine built on Leonardo's model that actually worked. A and B are the eye

† 81 r.   ‡ p. 78.

and the source of light, and are in long slots ; C is the centre of the circle ; P slides up and down a slot till I is on the circle of centre C. $INMP$ is a rhombus so that PI produced bisects the angle $\angle AIB$.

§ 5. The inscription of regular polygons

Leonardo was much interested in anything which was connected with drawing. He consequently, and quite naturally, gave some attention to geometrical constructions. Marcolongo does not consider these worth attention, but Cantor was of a different opinion, and gives them not a little attention in Cantor[1].†

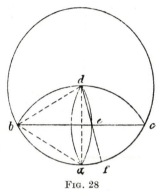

FIG. 28

It is to be noted at the outset that Leonardo used a compass that was capable of but one opening. This restriction is usually assigned to Abul Wafa, but it is not certain that Pappus did not have the same idea. We find specifically in Leonardo[3]‡ 'affare una linea curva divisa in parti dispari e eguali come appare in *abc* con un solo aprire del seste'. This raises the question of whether Leonardo was familiar with Abul Wafa's work. I shall presently give his construction for the regular hexagon which is essentially that of his Arab predecessor. However, this particular trick is found in the work of other writers ; the evidence does not seem to me convincing.

Let us first find his determination of an angle of 15°. I take Cantor's drawing as clearer than Leonardo's original, Fig. 28. The centre of the circle is d, the arcs bd, da, ac, cd are each 60°. We draw bc bisecting arc da in e. The arc de is 30° so $\angle adf = 15°$. We can now easily inscribe a regular polygon of three, four, or six sides. Leonardo made various attempts at inscribing a regular polygon of five sides. Some of these were so obviously incorrect that he himself labelled them *falso*. This was an old problem ; there is a correct solution in Euclid, iv. 11, and another is given by Abul Wafa.§ The fact that Leonardo gives his own incorrect solution does not surely prove that he had not seen Abul Wafa's work ; he surely must have been familiar with Euclid's construction. Perhaps he did not think these easy to execute in practice, and was seeking for something simpler. His best attempt here is, put in slightly different form, 'Given a line segment, to construct a regular pentagon with this as one side.' The construction will come immediately if we can find the radius of the circle in which such a pentagon will be

† q.v., vol. ii, pp. 295 ff.   ‡   15 v.
§ Woepcke (q.v.), p. 327.

inscribed.† The given segment is da; we find p and r, the intersections of equal circles with radii da and ad; s is the middle point of da, which we divide into eight equal parts. We draw pg parallel to ad and equal

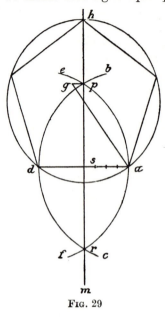

FIG. 29

to $ad/8$. Then ag meets ps in the centre of the circle required. This is, as a matter of fact, a good approximate construction as it makes sin $36° = 0·585$ instead of $0·587$.

It is interesting to speculate as to just what led Leonardo to make this particular construction. p is the centre of the circle where ad is the side of a regular hexagon. If it is to be the side of a pentagon, the radius must be shortened. Leonardo would have us go one fifth of the way down to s; did the fact that the pentagon has five sides influence his choice? The scheme of moving the centre up and down appealed to him. He knew of a correct construction of the regular octagon, but in Leonardo[3]‡ we have an incorrect one, Fig. 30. He takes the distance pC one-third of a radius above p, and says that C is the required centre. Cantor§ points out that the broken line $Cpa = \frac{8}{6}ap$, the coefficient is the ratio of the numbers of sides of the octagon and hexagon. I should put this down as altogether too fantastic to be the real reason for the choice of this method, were it not repeated in Leonardo[3],‖ where it is a question of inscribing a regular nonagon. ng (Fig. 30, right) is one-half the radius, so that the broken line

$$gna = \tfrac{3}{2}na = \tfrac{9}{6}na.$$

Now Cantor†† divides Leonardo's constructions into three classes. The first class are stigmatized with the adjective *falso*. Of these nothing need be said. The second class are not qualified by any adjective; Cantor thinks that Leonardo considered them good enough, even if not geometrically perfect. To the constructions of the third class the adverb *apunto*, i.e. 'exactly', was attached. Cantor interprets this as meaning that Leonardo looked upon them as mathematically perfect. But this adjective is attached to the construction of the nonagon which I have just given; it is hard to believe that Leonardo really believed in perfection here. It seems to me far more likely that he tried the thing experimentally and found the construction quite as accurate as he believed

† Leonardo[3], B. 13 v.   ‡ B. 14 r.   § Cantor[1] (q.v.), vol. ii, p. 299.
‖ B. 29 r.   †† Ibid., p. 300.

necessary. The same thing applies to his construction of the regular heptagon.† He takes as a side of an inscribed heptagon the altitude of one of the six equilateral triangles into which the inscribed hexagon is divided by radii to the corners, or, more simply, half the side of the

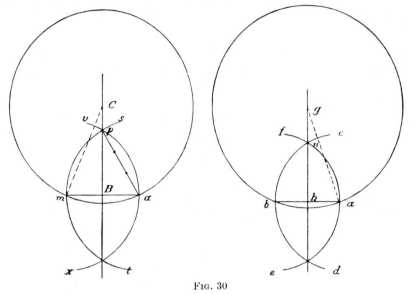

FIG. 30

inscribed equilateral triangle. This was not correct, nor was it original; it was called the Indian rule, and was given by Abul Wafa.‡

§ 6. Centres of gravity

One other purely mathematical topic attracted Leonardo's attention, that of centres of gravity. For this there were many reasons. Leonardo was an enthusiastic if not, perhaps, a very deep student of the works of Archimedes. This writer published two documents bearing on the equilibrium of plane figures, and there seems reason to believe that he had other writings on similar subjects which are now lost. Moreover, mechanical problems of this sort were naturally intriguing to Leonardo anyway, and the centre of gravity is fundamental in statics. Lastly Leonardo, as an artist, was interested in questions of balance; he treated them in his work on painting.

I notice first that Leonardo, following Albert of Saxony, seems to have done his best to bedevil the subject by introducing three different sorts of centres of gravity. We find him writing: 'In ogni figura pon-derosa si trova essere tre cientri de quali luno e cientro della gravità naturale, il 2° della gravità accidentale, il 3° del magnitudo del

† Leonardo², B. 28 r.   ‡ Cantor¹, vol. ii, p. 83, and Woepcke (q.v.), p. 320.

corpo.'† These three terms appear again in other places. Marcolongo interprets *cientro della gravità naturale* to mean the centre of mass, while *magnitudo* would give the centre if the body were homogeneous. What *accidentale* means I do not know, but it seems to come in only when extraneous weights are added, as in the case where the weight of a lever has been added to those of other weights applied. Perhaps Leonardo himself did not have the distinctions very clearly in mind.

Archimedes gave two excellent proofs of the theorem that the centre of gravity of a homogeneous triangle is the point of concurrence of the medians. Leonardo was cognizant of this fact: 'Ogni triangola a il cientro della sua gravità nella intersegatione delle linie che si partano dalli angoli e terminando nella metà della opposita base.'‡

Leonardo does not give a proof, but he would naturally assume that a triangle would balance on a knife-edge that ran along a median. When it comes to finding the centre of gravity of an isosceles trapezoid, he gives an incorrect answer in Leonardo[4], 3 r., but a perfectly correct one in 17 v. First we connect the middle points of the parallel sides. Then we divide into two triangles by means of a diagonal, and connect the centres of gravity of the two. The centre of gravity sought is on each of these lines. Leonardo pays some attention to the centre of gravity of a pentagon, dividing it into triangles and trapezoids. At this point Marcolongo seems to be confused, for he writes: 'Riferendosi poi per il resto alla 5 di Archimede.'§ Now I find in Leonardo[3], 17 v, no reference to Archimedes. The point is important, as bearing on the question of whether Leonardo had ever seen Archimedes' work on centres of gravity of plane figures. If he had, why did he sometimes give incorrect answers to questions which Archimedes answered correctly?

Near the close of Leonardo[4]‖ is an attempt to find the centre of gravity of a semicircle. He divides the figure into eight equal sectors, which he handles as if they were isosceles triangles. The result is fairly accurate, but it shows that he had never heard of Pappus' theorem about the volume generated by rotating a closed plane figure about a line which does not cut it.

I have so far pointed out the weak points in Leonardo's studies in centres of gravity. If there were nothing else we should do better to omit the subject entirely. But there is something else, and it is, in fact, his one great contribution to geometry: he determines the centre of gravity of a tetrahedron. Here is his statement: 'Il cientro di gravità del corpo di 4 basi triangolari sia nella segatione de sui assi, e sarà nella 4ª parte della sua lunghezza.'††

We are brought face to face with the question of how Leonardo

---

† Leonardo[4], 72 v.   ‡ ibid., 16 v.   § Marcolongo, p. 196.
‖ 215 r.   †† Leonardo[4], 193 v.

discovered this. There is no trace of a proof, at least in any writing of his that has come down to us. If Archimedes really wrote a now lost treatise on centres of gravity it seems almost certain that it contained this theorem, and carefully proved, but this is mere speculation. Some commentators have become excited over the point, thus Libri:

'Dans le volume F (f. 51) il détermine le centre de gravité comme cela est, en effet, au quart de la hauteur de la droite qui joint le sommet au centre de gravité de la base; la figure qui accompagne sa note prouve que Léonard décomposait la pyramide en plans parallèles à la base comme on fait à présent.'[†]

This caused an outbreak on the part of Duhem:

'Libri a écrit avec son inexactitude habituelle "La figure qui accompagne" etc. En réalité les deux figures dessinées par Léonard ne portent aucune trace de cette décomposition. Léonard a simplement tracé les médianes en chacune des diverses bases du tétraèdre, et les lignes qui joignent chaque sommet au point de concours des médianes de la face opposée.'[‡]

This explicit statement of Duhem's is certainly correct, but even if the figures do not show it, Leonardo may well have divided up his tetrahedra into thin slices parallel to the base; I personally believe he did so. Here are my reasons.

Leonardo makes great play of what he calls the 'axis' of any pyramid; it is the line joining the vertex to the centre of gravity of the base; note that I say 'any pyramid', not merely tetrahedron. Here is the way that the tetrahedron theorem is phrased:

'Il centro di gravità piramidale è nel quarto de suo assis di verso la base, e se dividerà l'assis per 4 eguali e intersegherai due a due li assi di talc piramide, tale intersegatione verrà nel predetto quarto.'[§]

Or again:

'Di ogni piramide tonda triangola, quadratica, o di quanti lati sia, il centro di gravità è nella 4 parte della sua assis vicina alla base.'[‖]

I should like to point out next that by proposition 8 of the first book of Archimedes' brochure on the equilibrium of plane figures, assuming that Leonardo was familiar with it, the centre of gravity of a figure composed of two parts is shown to be on the line joining the centres of gravity of the individual parts, and the same will hold in three dimensions, and when the number of parts is greater than three, provided their centres of gravity are all collinear. Furthermore, by Archimedes' fifth postulate in the same work, the centres of gravity of similar figures are similarly placed. It will follow from this that if we take any

[†] Libri (q.v.), vol. iii, p. 41.   [‡] Duhem (q.v.), vol. ii, p. 75.
[§] Marcolongo[1], p. 200 (the reference given there to Leonardo is incorrect).
[‖] Leonardo[4], 218 v.

number of thin sections of a pyramid parallel to the base, their centres of gravity will all lie on a line through the vertex, and the centre of gravity of the whole pyramid will lie on this line. If the pyramid be triangular, if we draw two such 'axes' the figure will show at once they divide one another in the ratio 1:3. The centre of gravity of a triangular pyramid is thus one-quarter of the way up an axis. Hence, if we divide any pyramid into triangular ones, its centre of gravity will be one-quarter of the distance up the axis, for the centres of gravity of all the partial pyramids are in a plane one-quarter of the way up from the base. It seems to me altogether likely that Leonardo reached his theorem in this fashion, especially because Archimedes probably proceeded along similar lines. We have learnt in recent times that although Archimedes proved theorems in measurement by the most rigorous use of the method of exhaustion, he frequently first discovered the facts by sub-dividing into thin slices.[†] Why might not Leonardo have done just the same thing, omitting the rigorous demonstration ? Our whole study of Leonardo suggests that he was only faintly interested in rigorous mathematical demonstration; what he cared for was mathematical work that gave the answer with all of the accuracy needed in any practical case. The underlying philosophy of mathematics meant very little to him, the concrete results meant a great deal.

† Heath (q.v.), vol. ii, p. 21.

CHAPTER V

ALBRECHT DÜRER

§ 1. Spirals and helices

IT is a noteworthy fact in geometry that the study of figures in three or more dimensions has lagged very far behind that of figures which are in one plane. We have already mentioned on p. 12 that this was a source of regret to Plato. There was really a deep underlying reason for this, even if Plato were unaware of the fact. It can be shown that the geometry of the plane is not only simpler than that of higher spaces, but actually richer in interesting results. Such matters are not, however, our present concern; historically the lag of solid geometry behind plane geometry has persisted from Plato's time to our own.

The idea has come at different times to different mathematicians to make three-dimensional geometry more vivid by representing space figures in the plane. We cannot represent all of three-space in one plane, for obvious reasons, but suppose that we try to represent in two planes simultaneously, and then superpose one plane on the other. A point in three-space will then be represented by two points in the plane. Clearly these must be connected in some definite way, for the points in three-space depend on three independent parameters, while two points in the plane depend on four. There are various methods of overcoming this simple difficulty.

The most usual procedure is called 'descriptive geometry'. The credit for discovering it is generally ascribed to that prince of teachers, Gaspard Monge. We project our space figure orthogonally and vertically on a horizontal plane, and horizontally on a vertical plane, then we rotate one plane about the line of intersection of the two until it coincides with the other. A point in three dimensions is thus represented by two points in the plane, one vertically above the other. In order to put this in more concrete mathematical form, let us assume that we have in space an English rectangular Cartesian system of axes. The plane of the paper, the vertical plane, shall be the (Y, Z) plane, the positive Y-axis going horizontally to the right. The positive X-axis shall come directly towards the observer. This (X, Y) plane is then rotated around the intersection downwards until the X-axis takes the $-Z$ direction. The point (X, Y, Z) is thus represented by the pair of points (Y, Z), $(Y, -X)$, that is, if the line of intersection of the two planes is taken as the Y-axis.

Let us give Monge all possible credit for discovering this simple and useful process; the credit for actual discovery belongs to the great artist of Nuremberg whose name appears at the head of the chapter. The first to point out this interesting fact seems to have been Amedeo (q.v.). It must not be imagined that Dürer, in writing Dürer[1], imagined that he was composing a complete treatise on geometrical science. He wished to give a sufficient number of definitions and first principles so that his reader could perform easily certain drafting operations. His

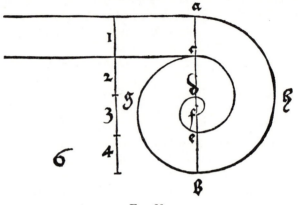

Fig. 31

definitions are decidedly sketchy. A point is something which has neither length, breadth, nor thickness, a line is the path of a moving point; *Relinquit post se vestigium.*

Dürer was particularly interested in helical space curves. The projection of such a curve on a plane, perpendicular to the cylinder on which it lies, is a spiral, so he begins with plane spirals.† The simplest, which I reproduce in Fig. 31, is drawn with the aid of a compass alone. We make a series of semicircles, whose diameters lie along a certain line, the curves lying alternately above and below it, each sharing an end with the preceding and with the following curve. We have here a continuous curve with a continuously turning tangent, but there is a disquieting set of discontinuities in the curvature. Dürer must have been worried by this, for he proceeded to show a variety of methods of shortening the radius vector as the latter turns around. The whole story is developed in Dürer[1],‡ but instead of starting with the simple case of the circular helix, which comes later, he takes a decidedly complicated curve. This does not lie on a cylinder, nor yet on a cone, but makes one turn around a cylinder and one around a cone, which stands on top of it. The azimuth θ shall run from $0°$ to $120°$, the height shall be given by

$$z = a \tan \frac{\theta}{24}.$$

† Dürer[1], p. 4 f.   ‡ pp. 13, 14.

For the first turn, the horizontal projection on the (X, Y) plane is $r = b$; for the second turn we have the much more complicated form

$$r = b \left[\frac{\tan 30° - \tan \dfrac{\theta}{24}}{\tan 30° - \tan 15°} \right].$$

GEOMETRIAE   LIB.   ALBERTI DVRERI

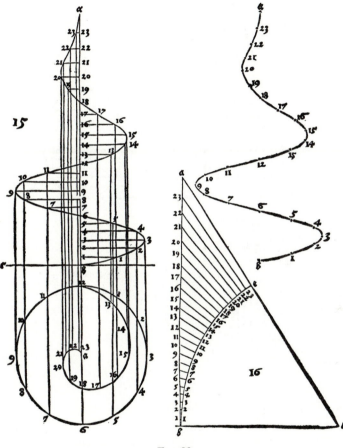

FIG. 32

The horizontal projection on the (Y, Z) plane is

$$y = r \sin \theta.$$

Dürer undertakes something much simpler when he passes to the conic sections. We show in Fig. 33 the projections of the ellipse. The projection on the (Y, Z) plane is a line segment bounded by two sloping lines.

This is divided into twelve equal parts, and through the points of
division vertical lines are drawn and numbered. Horizontal lines also are
drawn through the points of division ; the segments determined on them
by the sloping lines will be the diameters of the circular sections which

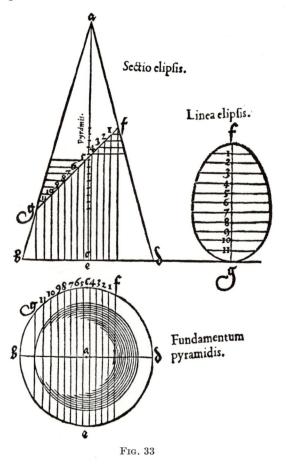

Sectio elipſis.

Linea elipſis.

Fundamentum
pyramidis.

Fɪɢ. 33

horizontal planes cut from the cone. We draw in the (X, Y) plane a
series of circles with these diameters, and when that is folded down to
lie on the (Y, Z) plane we have a series of concentric circles just below
the figure in the (Y, Z) plane. Where each of these circles meets the
vertical line with the same number will be the folded-down projection
of a point of the ellipse. We have in this way a rather unshapely projection of the original curve. However, Dürer does not draw it in, but
rather constructs the curve itself. The original line segment has the
length of the major axis. This we set upright to the right of the figure,
and divide into twelve equal parts. Through each point of division we

draw a double ordinate equal to the diameter of the corresponding horizontal circle.

Before leaving this part of the subject I will describe an even more

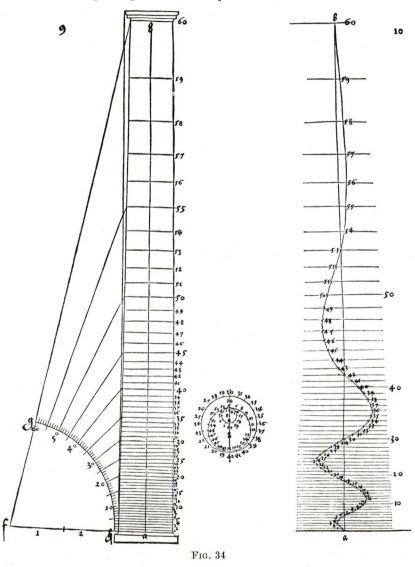

Fig. 34

complicated sort of twisting space-curve. This lies on three cylinders of revolution, each tangent to the next along a vertical element. The intersections with the (X, Y) plane, the vertical projection of the curve, are three tangent circles, which are treated as two spirals of the sort mentioned on p. 62. The middle circle is tangent to the two others at points where $Y = 0$.

We start on the smallest circle at a point, not a point of contact, where $Y = 0$, and divide into six equal parts numbered 1 to 6; we pass then to the middle circle, continuing around in the same sense of rotation and in a half-turn take twelve equal parts numbered 7 to 18. Continuing always to turn in the same sense we pass to the largest circle and make a complete circuit with equal parts numbered 19 to 42, then around the other half of the middle circle with parts numbered 43 to 54, and lastly on the smallest circle with parts numbered 55 to 60. The space curve is wound in this order, so that it goes around each imaginary cylinder once. To write its equations we suppose r_1, r_2, r_3, the radii of the three circles, the azimuth θ. We also choose an arbitrary angle α which looks like $60°$ in the figure, and the height of the column h. If, then, the point numbered n have the azimuth θ, which will depend on whether the semicircle is divided into six parts or twelve, we have

$$y = r_i \sin \theta; \qquad z = \frac{h \tan(n\alpha/60)}{\tan \alpha}.$$

I find it hard to see what led Dürer to choose this complicated curve.

§ 2. Problems in one plane

Dürer paid some attention to classical geometrical problems as well as to plane curves of his own devising. Here is his construction for a heart-shaped curve of the fourth order, drawn with a double ruler so constructed that one part makes twice the angle with the horizontal that the other does

$$ab = r; \qquad bc = p;$$

$$x = r \cos \theta + p \cos 2\theta; \qquad y = r \sin \theta + p \sin 2\theta;$$

$$x^2 + y^2 = r^2 + p^2 + 2rp \cos \theta;$$

$$2p(x+p) = [x^2 + y^2 - (r^2 + p^2)]\left[1 + \frac{1}{r^2}\{x^2 + y^2 - (r^2 + p^2)\}\right].$$

He makes various attempts to inscribe regular polygons in a circle, probably realizing that what he gives are only approximate constructions. In Cantor[1], vol. ii, eight pages are given to describing this part of Dürer's work to the exclusion of other parts of greater significance. Thus we read:† 'Albrecht Dürer ist der erste, welcher die Nährungskonstruktionen mit vollem Bewusstsein ausgeführt hat.' This seems to me a difficult thesis to defend, especially in view of Leonardo's

† p. 465.

work. To inscribe a regular heptagon Dürer takes one-half of a side
of an inscribed equilateral triangle, a classical construction we have

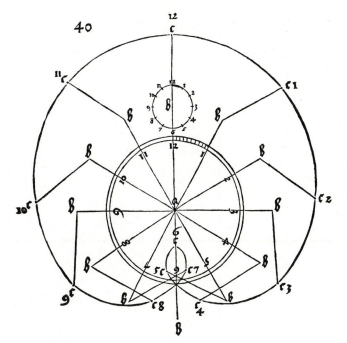

FIG. 35

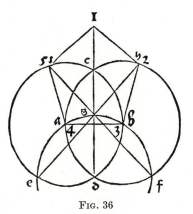

FIG. 36

mentioned already on p. 57. For the pentagon there are two attempts:
here is the better one. We have, given the side, to find the circum-
scribed circle. α, β, δ are the centres of three equal circles. Find 2 and 5,
the points where the lines from e and f to the middle point of the arc $\alpha\beta$

meet the circles about α and β. He takes 5α, $\alpha\beta$, $\beta2$ as three sides of the pentagon. This leads to the approximation

$$\sin 27° = 2 \sin 60° \sin 15°,$$
$$0\cdot454 = 0\cdot448.$$

His approximation for the ratio of circumference to diameter is $\pi = 100/32$.

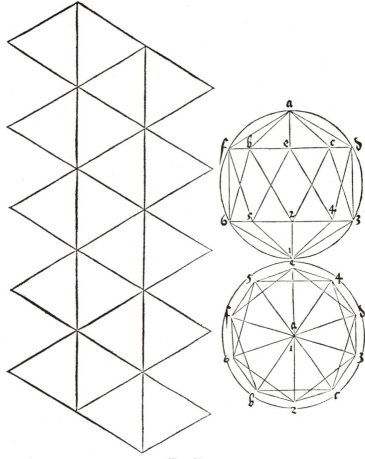

Fig. 37

Amusing construction appears at the beginning of his Book IV to construct regular solids by paper folding. This is the usual procedure taught to-day in our schools; so far as I can make out it is original with Dürer.† I reproduce his picture for the icosahedron in Fig. 37. For duplicating the cube he uses what is essentially that of Sporus, given

† Cantor[1], vol. ii, p. 466.

by Eutocius in his commentary on Archimedes and reproduced in Valla (q.v.).

§ 3. Descriptive geometry

It is a strange fact that Dürer's most original geometrical work does not appear in Dürer[1] but in Dürer[2], which was composed subsequently. This contains a fantastic number of measurements of human bodies,

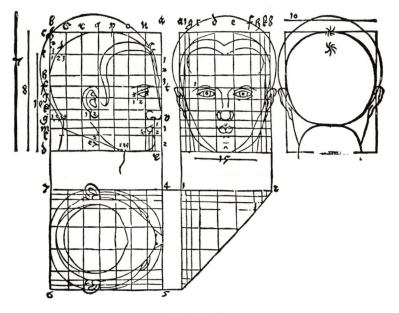

FIG. 38, 39

male and female, tall and short, stout and thin, straight and bent, with projections seen from front, back, or sides. But Dürer is sometimes anxious to give not two orthogonal projections but three, on three mutually perpendicular planes, which is something Monge did not undertake. We project on the (Y, Z), (X, Y), and (X, Z) planes. The (X, Y) plane is then swung down as before until it becomes the $(-Z, Y)$ plane, and the (Z, X) plane is rotated until it becomes the $(Z, -Y)$ plane. Dürer perceives that these three are not independent one of another, but if we know two projections we can construct the third. Corresponding points on the (Y, Z) and the swung-down (X, Y) plane are on the same vertical lines, corresponding points on the (Y, Z) and the swung (X, Z) planes are on the same horizontal lines. The coordinate which was originally X becomes in one case $-Z$, and in the other $-Y$. The relation between the two rotated planes is thus an interchange of Y and Z, and this we perform by drawing a line, which Dürer calls a 'transferent', which makes an angle of $45°$ with the horizontal.

Horizontal lines of the rotated (X, Y) plane are brought to meet this, then rotated through an angle of 90° to become the vertical lines of the rotated (X, Z) plane. This appears very clearly in Figs. 38 and 39.

It is fair to say that Dürer does not seem to have seen anything of

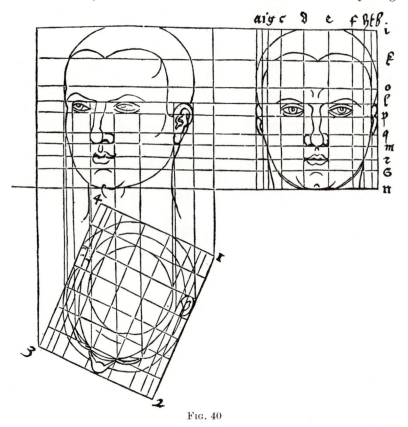

Fig. 40

geometric interest in this, merely a way of passing from a side view of a head to a front view. This suggested to him that we need not turn through an angle of 90° but through any other angle we please. If we turn the head about a vertical axis, the projection in the (X, Y) plane will be turned through the same angle. If we turn the figure in the rotated (X, Y) plane through any convenient angle, then erect perpendiculars, they will meet the horizontal lines in the rotated (X, Z) plane in points giving the projection on the (Y, Z) plane of the head turned about a vertical axis. We see this in Fig. 40 and several times later. One wonders whether Dürer was aware of the fact that any rotation is the product of two reflections. This might lead to a further use of transferents.

CHAPTER VI

JOHN NAPIER, BARON OF MERCHISTON

§ 1. Logarithms

WE must, I think, accept as a fact that most of the contributions to mathematics made by men whom I class as great amateurs may be beautiful or attractive or ingenious, but seldom indeed are they what we may call practically useful. Perhaps Leonardo's studies of war machines, or better his work in hydraulics, perhaps Pietro's studies of perspective or Dürer's invention of descriptive geometry, should not be condemned as unpractical, but the general statement remains true. But now we come to a glaring exception, in the work of the theologically contentious Scot whose name appears above. The inventor of logarithms will always be classed as one of those who have benefited mankind in a directly practical way, even if we do not include his attacks on the Church of Rome, or his invention of secret weapons of war as striking benefactions.

John Napier himself failed to appreciate the full importance of his invention of logarithms; he believed that his interpretation of the Apocalypse of St. John was his greatest service to mankind. Perhaps we should not judge him too hardly for this: a far greater than he, Sir Isaac Newton, is said to have had an exactly similar illusion. But Napier had a strong desire to do something to shorten the labour of astronomical calculation and other applications of trigonometry, subjects in which he was deeply interested. He was very much aware that addition and subtraction are less laborious than multiplication and division, especially where large numbers are involved, and he set himself to the useful task of devising a method of avoiding the more laborious techniques. Lord Moulton, in an interesting article on the invention of logarithms which appears in Napier[2], points out that Napier's logarithms were originally logarithms of sines, and the formula

$$\sin A \sin B = \tfrac{1}{2}[\cos(A-B)-\cos(A+B)]$$

may have given him the first hint of substituting addition and subtraction for multiplication and division. I cannot feel that this point is well taken. We shall see that he began with a comparison of arithmetical and geometrical series, even though his first logarithms were merely those of sines. But this relation between the two series was known to Archimedes† and appears in his Sand Reckoner. Napier received much of his education on the continent of Europe, and is believed to

† Archimedes[2], p. 229.

have been familiar with the work of Stifel. This writer is perfectly explicit:

'Additio in Arithmeticis progressionibus respondet multiplicationi in Geometricis.

'Subtractio in Arithmeticis respondet in Geometricis divisioni.

'Multiplicatio simplex (id est, numeri in numerum) quae fit in Arithmeticis respondet multiplicationi in se quae fit in Geometricis ... ut in superiore exemplo 5.11.17 triplatio medii facit quantum additio omnium trium, sic hic 4.6.9 multiplicatio cubica medii facit quantum tres inter se.

'Divisio in Arithmeticis progressionibus respondet extractionibus radicum in progressionibus Geometricis.'†

How could a man who had written this fail to discover logarithms as exponents?—especially when he wrote such arrays as

$$0 \quad 1 \quad 2 \quad 3 \quad 4 \quad 5$$
$$1 \quad 2 \quad 4 \quad 8 \quad 16 \quad 32$$

Smith says that he went further and considered negative powers.‡ I have not been able to verify this in Stifel's work, but there is an even more striking array in Clavius, where we find

$$-7 \; -6 \; -5 \; -4 \; -3 \; -2 \; -1 \; \; 0 \; 1 \; 2 \; 3 \quad 4 \quad 5 \quad 6 \quad 7$$
$$\frac{1}{128} \; \frac{1}{64} \; \frac{1}{32} \; \frac{1}{16} \; \frac{1}{8} \; \frac{1}{4} \; \frac{1}{2} \quad 1 \; 2 \; 4 \; 8 \; 16 \; 32 \; 64 \; 128$$

with the comment: 'Nam et hic superioris progressionis numeri idem faciunt additione et subtractione quod inferioris progressionis numeri multiplicatione atque divisione.'§ I do not know whether Napier was familiar with this. If he was, it is hard to see why he set up his logarithms in the clumsy fashion which I shall describe presently; it is equally surprising that neither of these writers anticipated his discovery.

Napier's writings on logarithms are two in number. First we have the *Mirifici logarithmorum canonis descriptio* (Edinburgh, 1614). A translation was published by Wright in 1616. The *Constructio canonis logarithmorum* was published in Edinburgh in 1619; a translation by Macdonald appeared in the same city in 1889. In dealing with Napier's logarithms I shall generally follow this latter. Let us now see how he goes to work.

Napier starts to set up, side by side, arithmetical and geometrical series, pointing out that the construction of the former is usually much the easier. If a geometrical series is a descending one, the process consists in subtracting each time a certain proportion of what is left. If

---

† Stifel (q.v.), pp. 35, 36.   ‡ See Napier[2], p. 86.
§ Clavius (q.v.), vol. ii, algebraic part, p. 16.

this last factor of proportionality is a negative power of 10, we take from
what we have what appears when we have moved the decimal point a
certain number of places to the right. Napier begins like this

$$
\begin{array}{r}
10000000{\cdot}0000000 \\
1{\cdot}0000000 \\
\hline
9999999{\cdot}0000000 \\
{\cdot}9999999 \\
\hline
9999998{\cdot}0000001 \\
{\cdot}9999998 \\
\hline
9999997{\cdot}0000003\dagger
\end{array}
$$

At this point I must speak of Napier's use of the decimal point and
of decimal fractions, each of which he manipulated with skill. Ball says
with regard to the fractions:

'In Napier's posthumous construction published in 1619 it is defined and
used systematically as an operative form, and as this work was written after
consultation with Briggs about 1615–16 and probably revised by the latter
before it was issued, I think it confirms the view that the invention was due
to Briggs and was communicated by him to Napier. At any rate it was not
employed as an operative form by Napier in 1617.'‡

If Ball means by this that the notation for decimal fractions was
invented by Briggs he is certainly in error. The notation had been
passing through a long process of evolution, the decimal point was
used by Pellos in 1493. Here is a more recent view:

'It is unquestionably true that the invention of logarithms had more to do
with the use of decimal fractions than any other influence. When Napier
published his tables in 1614 he made no explicit use of decimal fractions, the
sine and the logarithm being each a line of so many units. In the 1616
translation of the work, Edward Wright used the decimal point.'§

The most recent view that I have seen is that of Gandz, who says that
the systematic use of decimal fractions, and of integral and fractional
exponents, was first introduced in the middle of the fourteenth century
by Immanuel Bonfils.‖ I have not been able to verify this statement,
but it seems to me unlikely, even if this writer first discovered these
useful notations, that their use was widely known in Napier's time, or
that he copied from him.

I must now explain Napier's actual line of reasoning. Instead of
following the natural arithmetical route suggested by the relation of
arithmetical and geometrical progressions, he introduced geometrical

† Napier[1], p. 13.   ‡ Ball (q.v.), p. 204.
§ Smith (q.v.), vol. ii, pp. 244 and 338.   ‖ Gandz (q.v.), p. 273.

considerations which seem to me to have complicated the matter enormously. His sines, for I must repeat that it is only the logarithms of sines with which he is concerned, are not supposed to decrease arithmetically, but in a geometrical series. This led him to the idea of rates. He imagines two moving points. The first goes at a uniform velocity, say one unit of distance for each unit of time. The second, starting at a distance r from its goal with the same velocity as the first, goes at a decreasing speed, its velocity during each interval of time being proportional to its distance from its ultimate goal at the beginning of the interval. Its distance from that goal at the end of any time-interval will be its distance at the beginning less $1/r$ of the same. The distances from the goal are, then,

$$ r, \quad r\left(1-\frac{1}{r}\right), \quad r\left(1-\frac{1}{r}\right)^2, \quad ..., \quad r\left(1-\frac{1}{r}\right)^y. \tag{1} $$

If thus the distances from the start in the first series are

$$ 0, \quad 1, \quad 2, \quad ..., \quad y, $$

we see in this way how the numbers in one series correspond to the exponents in the other. Here is Napier's own statement: 'The logarithm of a given sine is that number which has increased arithmetically with the same velocity throughout as that with which radius began to decrease geometrically and in the same time as radius has decreased geometrically to the given sine.'[†]

Before leaving these equations I will point out:

if
$$ x = r\left(1-\frac{1}{r}\right)^y \quad \text{and} \quad x' = r\left(1-\frac{1}{r}\right)^{y'}, \tag{2} $$
$$ \frac{x'}{x} = \left(1-\frac{1}{r}\right)^{y'-y}. $$

So that if the x's follow a geometrical series, the y's will follow an arithmetical series, and this is the heart of the whole matter. The number r which he calls the 'whole sine' he takes, following Regiomontanus[‡]
$$ r = 10^7. \tag{3} $$

I must now quote the opinion of Lord Moulton that there were three successive stages in the development of Napier's thought. The first stage was concerned with the construction of a decreasing geometric series.

'In the second stage he found himself repeatedly deducting from a number say its ten millionth part, and the continued multiplication by a factor gave way to taking away one and the same aliquot part of the number arrived at by the preceding operation. This naturally led him to pass in his thought

† Napier[1], p. 19.   ‡ Smith (q.v.), vol. ii, p. 242.

from figures representing quantities arithmetically to the geometrical representation of the quantity by a line, so that the repeated operation was
perfectly represented by repeatedly cutting off one and the same fraction of
the line operated on. . . .

'The concentration of Napier's thought on groups of operations identified
with equal lengths on the logarithmic line, prepared him for what I view as
the third and most interesting stage of the discovery. So long as the effect
of the group in the reduction it produced was the same, what mattered it
whether it was made out of a large number of small reductions, or a smaller
number of large ones? In both cases the principle that the logarithms of
proportional quantities are equidifferent would apply. It would only mean
that the determinate moments during which the point kept its velocity would
be shortened, and the changes of velocity would come more frequently. From
this he, no doubt, gradually passed into a stage of contemplating the changes
as taking place so frequently that it might be said that at each instant the
moving point possessed the exact velocity that it should have were it starting
to move for a determinate moment, i.e. its velocity is equal to its distance
from the end of the line.'[†]

This is certainly logical and the idea of velocity is there, but I am
not convinced that Napier had really the difficult idea of an instantaneous velocity which was so baffling to Newton and the other early
writers on the calculus. Napier himself says: 'whence a geometrically
moving point approaching a fixed one has its velocities proportionate
to its distances from the fixed one'.[‡] The fact that the word 'velocities'
appears in the plural suggests to me that he had in mind successive
steps rather than a continuous change.

It is fair to say that what is involved here is a very small point, of
theoretical interest only; the practical result is the same whether we
adopt one hypothesis or the other. We get from equation (2)

$$x = r\left(1 - \frac{1}{r}\right)^y,$$

$$\log_e x - \log_e r = y \log_e\left(1 - \frac{1}{r}\right),$$

$$\log_N x = y = \frac{\log_e x - \log_e r}{\log_e\left(1 - \frac{1}{r}\right)}.$$

Developing the denominator in power series, and multiplying above
and below by $-r$,

$$\log_N x = \frac{r(\log_e r - \log_e x)}{1 + \frac{1}{2r} + \frac{1}{3r^2} + \cdots}. \tag{4}$$

If, on the other hand, we adopt the hypothesis of instantaneous velocity,

[†] Napier[2], pp. 11 and 14.   [‡] Napier[1], p. 18.

and give to the first point the velocity 1, then when the second point
is at a distance x from its goal, its velocity is $-x/r$. As the logarithm
is the distance covered by the first point, which is equal to the time it
has been moving, and this is the time the second point has moved, we
have

$$\log_N x = -r \int_r^x \frac{dx}{x} = r(\log_e r - \log_e x). \qquad (5)$$

This is the value usually given for Napier's logarithm. The ratio of the
two, in view of (3), is

$$1 + \frac{1}{2 \cdot 10^7} + \frac{1}{3 \cdot 10^{14}} + \dots.$$

It is now time to show how Napier calculated his logarithms.† He
starts with

$$r, \quad r\left(1-\frac{1}{r}\right), \quad r\left(1-\frac{1}{r}\right)^2, \quad \dots, \quad r\left(1-\frac{1}{r}\right)^{100} = r\left(1-\frac{1}{10^5}\right).$$

The last equation is only approximately correct, it amounts to writing

$$\left(1-\frac{1}{r}\right)^n = 1-\frac{n}{r}.$$

The logarithms of these numbers are 0, 1, 2,..., 100.

We now write a second series:

$$r, \quad r\left(1-\frac{1}{10^5}\right), \quad r\left(1-\frac{1}{10^5}\right)^2, \quad \dots, \quad r\left(1-\frac{1}{10^5}\right)^{50} = r\left(1-\frac{1}{2000}\right).$$

These also are a geometrical series, so that their logarithms from the
arithmetical series are 0, 100, 200,..., 5,000.

The third and final table contains sixty-nine columns; the first is

$$r, \quad r\left(1-\frac{1}{2000}\right), \quad r\left(1-\frac{1}{2000}\right)^2, \quad \dots, \quad r\left(1-\frac{1}{2000}\right)^{20} = r\left(1-\frac{1}{100}\right).$$

The logarithms of these numbers are

$$0, \quad 5{,}000, \quad 10{,}000, \quad \dots, \quad 100{,}000.$$

The logarithm of the number $r\left(1-\frac{1}{100}\right)^k$ will be $k00{,}000$; $k = 0, 1,$
2,..., 68. We have sixty-nine columns such as

$$r\left(1-\frac{1}{100}\right)^k, \quad r\left(1-\frac{1}{100}\right)^k\left(1-\frac{1}{2000}\right), \quad r\left(1-\frac{1}{100}\right)^k\left(1-\frac{1}{2000}\right)^2, \quad \dots.$$

The logarithm of the $(l+1)$th number here will be

$$k00{,}000 + 5{,}000l.$$

† Napier[1], pp. 13 ff.

Napier was perfectly well aware that in practice the logarithms, like the sines themselves, could only be given approximately. In such cases it is well to give limits of approximation. The sum of two positive numbers lies between the sum of their lower limits and that of their upper limits. When we have the difference between two positive numbers, which difference is supposed positive, that lies between the lower limit for the minuend less the upper limit for the subtrahend, and the upper limit for the minuend less the lower limit for the subtrahend. But Napier has a much shrewder approximation for sines and their logarithms. While the sines are decreasing in geometrical progression, the increase of the logarithms depends on the number of times the sines have decreased, not on the ratio of decrease.

Let the sine shrink from x to a. The logarithm increases at a steady rate, but the sine shrinks at a decreasing rate. At the start the rate of shrinkage of the sine is x/r times the rate of increase of the logarithm, at the close it is a/r times the rate of increase. The decrease of the sine is therefore less than it would have been had there been $\log a - \log x$ jumps each x/r times the rate of increase of the logarithm, i.e.

$$x-a < \frac{x}{r}(\log a - \log x).$$

Similarly

$$x-a > \frac{a}{r}(\log a - \log x),$$

$$\frac{r}{x}(x-a) < \log a - \log x < \frac{r}{a}(x-a).$$

In practice Napier uses

$$\log x = \log a \pm \frac{r}{a}(x-a). \tag{6}$$

It is time to introduce to our scene Henry Briggs, Professor of Geometry in Gresham College, London. The often quoted account is from William Lily.

'I will acquaint you with a memorable story related to me by Mr. John Marr, an excellent mathematician and geometrician whom I conceive you remember, he was a servant to King James, and Charles the First.

'At first when Lord Napier, or Lord Marchiston, made publick his Logarithms, Mr. Briggs, then reader of the Astronomy lecture at Gresham College in London was so surprized with admiration of them that he could have no quietness in himself, until he had seen that noble person the Lord Marchiston, whose only invention they were. He acquaints John Marr herewith, who goes to Scotland before Mr. Briggs, purposely to be there when two so learned persons should meet. Mr. Briggs appoints a certain day when to meet in Edinburgh, but failing thereof, the Lord Napier was doubtful he would not come. It happened one day as John Marr and Lord Napier were speaking of Mr. Briggs "Ah John", saith Marchiston, "Mr. Briggs will not come." At the very instant one knocks at the gate. John Marr hastened down, and it proved Mr. Briggs, to his great contentment. He brings Mr. Briggs to my Lord's chamber, where almost one quarter of an hour was spent each behold- ing the other with admiration, before one spoke: at last Mr. Briggs began: "My Lord, I have undertaken this long journey purposely to see your person and to know by what engine of wit or ingenuity, you first came to think of this most excellent help unto astronomy, viz. the Logarithms ? But my Lord, being by you found out, I wonder nobody else ever found it before, when now, being known, it appears so Easy." He was nobly entertained by Lord Napier, and every summer after this, during Lord Napier's being alive, this venerable man, Mr. Briggs, went to Scotland to visit him.'†

A good deal has been written about the relative contributions of Napier and Briggs to the method of logarithms. Perhaps the safest plan is to follow what Briggs says himself:

'That these logarithms differ from those which that illustrious man, the Baron of Merchiston, published in his Canon Merificus, must not surprise you. For I myself, when expounding their doctrine publicly in London to my audience in Gresham College, remarked that it would be more convenient that 0 be kept for the logarithm of the whole sine, (as in the Canon Mirificus) but that the logarithm of the tenth part of the whole sine, that is to say 5 degrees, 44 minutes and 21 seconds should be 10,000,000,000. And con- cerning that matter I wrote immediately to the author himself; and as soon as the season of the year and the vacation of my public duties of instruction permitted, I journeyed to Edinburgh, where being most hospitably received by him, I lingered for a whole month. But as we talked over the change in the logarithms he said that he had been for some time of the same opinion and had wished to accomplish it; he had, however, never published those he had already prepared, until he could construct more convenient ones if his affairs and his health would permit of it.   But he was of the opinion the change should be effected in this manner, that 0 should be the logarithm of unity, and 10,000,000,000 that of the whole sine, which I could not but admit was by far the most convenient.'‡

Let us look into this more closely, using modern notation. If we take the usual abbreviated form of definition for Napier's logarithms we have from (5)

$$\log_N x = r[\log_e r - \log_e x]. \tag{7}$$

Briggs substitutes 10^{10} for r, giving Napier's form altered by Briggs:

$$\log_N x = 10^{10}[\log_e 10^{10} - \log_e x].$$

Here we should have $\log_N 10^9 = 10^{10} \log_e 10$.

Briggs would prefer that this should be 10^{10}, which involves dividing

---

† Lily (q.v.), p. 235.   ‡ Gibson in Napier², p. 126.

by $\log_e 10$, and this again amounts to taking 10 as a basis for logarithms, so that in Briggs's scheme

$$\log_B x = 10^{10}[10 - \log_{10} x]. \tag{7}$$

Here again equal ratios will give equal logarithmic differences, the vital matter. Napier wished to retain this advantage, but from the equation

$$\frac{xy}{x} = \frac{y}{1}$$

we get $\qquad\qquad \log xy = \log x + \log y - \log 1.$

This would be much better if $\log 1 = 0$, the change he proposes. This will involve subtracting a constant from all logarithms, giving the final form
$$\log_N x = 10^9 \log_{10} x.$$

The factor 10^9 is unimportant.

It seems to me, to conclude, that Napier deserves the whole credit. He first saw the advantage of logarithms, put through a tremendous amount of work in calculating them by a highly ingenious method, for the development in power series was then scarcely known, thought out Briggs's improvement independently, and made a further improvement of his own. Our logarithm of to-day is essentially Napier's logarithm. What an accomplishment for a theologically inclined nobleman!

§ 2. Trigonometry

The admirable purpose of Napier's trigonometric work is to show the application of logarithms to trigonometric computation. At the outset the reader should be warned that Napier uses certain words in a sense that is different from what is current to-day. His logarithm table was for sines, so that when he speaks of the logarithm of an angle he means the logarithm of its sine, while the logarithm of a tangent is sometimes called 'differentialis'. His work in plane trigonometry begins with

'In rectangulo Logarithmus cruris est aequalis aggregato ex Logarithmo anguli oppositi+Logarithmo hypoteneusi

$$\log a = \log \sin A + \log c.$$

'In rectangulo Logarithmus cuiusvis cruris est aequalis aggregato ex differentiali oppositi anguli+Logarithmus reliqui cruris

$$\log a = \log \tan A + \log b.'$$

When he comes to oblique triangles he uses the law of sines whenever possible. When he has two sides and the included angle, or three sides given, he does not use the law of cosines, but proceeds otherwise. Here is his fifth proposition:

'In obliquangulis Logarithmus aggregati crurum, subductus a summa facto ex Logarithmo differentiae crurum, aequatur differentiali semi-aggregati suorum oppositorum angulorum relinquit differentiale semi-differentiae eorum.'[†]

This means:

$$\log(a-b)-\log(a+b) = \operatorname{logtan}\frac{A-B}{2} - \operatorname{logtan}\frac{A+B}{2}.$$

Which is rather better in the non-logarithmic form

$$\frac{a-b}{a+b} = \frac{\tan\frac{1}{2}(A-B)}{\tan\frac{1}{2}(A+B)}.$$

This we call the 'law of tangents'; it stems from Vieta.[‡]

Napier finds a wider scope for his methods when he passes to spherical trigonometry. The right and quadrantal triangles are treated together. He makes use of an ingenious device which is not, however, strictly new, as it was developed by Tarporley. This article I have not seen, but there is an adequate account in De Morgan (q.v.). The trick here is to take as the five 'circular parts' of a right spherical triangle the two legs and the complements of the hypotenuse and the two angles, other than the right. We find in this way a curious figure which was, apparently, in Tarporley's work. Begin with a right spherical triangle, and draw two arcs each of which is perpendicular to a leg and to the hypotenuse.[§] We have then a spherical pentagon and five right spherical triangles, each formed by a side of the pentagon, and two adjacent sides produced. It then turns out, strangely enough, that the same five values give the five circular parts of each triangle, they are merely arranged in different orders.

If we take three parts of a right spherical triangle, two must be next to one another. Then either the third is next to one of these, so that we have a part and the two next, or it is opposite to both, and we have a part and its two opposites. We thus get, in rather strange form, this fundamental rule of Napier's: 'Logarithmus intermediae aequatur differentialibus circumpositorum extremorum, seu Antilogarithmi oppositorum extremorum.'[||] This means: 'The logarithm of the sine of any part is equal to the sum of the logarithms of the tangents of the adjacent parts, or the sum of the logarithms of the cosines of the opposite parts.' I personally when a schoolboy learnt this in even shorter form. You were supposed to remember that you were dealing with the sine of a part, then you said: 'Tan ad, cos op.' The difficulty in practice consists in remembering in which case you take the given values and in which the complements.

When it comes to dealing with oblique spherical triangles, Napier

† Napier[3], pp. 25, 26.   ‡ Vieta (q.v.), p. 402.
§ Napier[3], p. 32. || Napier[3], p. 33.

shows considerable ingenuity in 'massaging' known formulae. I will not give many details here, merely referring to von Braunmühl† or Tropfke.‡

There is further trigonometric work in Napier[1]. The avowed object is to solve the oblique triangle without dividing it into right triangles. He is much impressed with the virtues of halved versed sines. For instance we have:

'*Given two sides and the contained angle, to find the third side.*

'From the half versed sine of the sum of the sides, subtract the half versed sine of their difference; multiply the remainder by the half versed sine of the contained angle, divide the product by the radius, to this add the half versed sine of the difference of the sides, and you will have the half versed sine of the required base.'§

If we take the radius as unity:

$$\left[\frac{1-\cos(a+b)}{2}-\frac{1-\cos(a-b)}{2}\right]\left[\frac{1-\cos C}{2}\right]+\frac{1-\cos(a-b)}{2}=\frac{1-\cos c}{2}.$$

This is rather clumsy, even if we have a table of haversines, as it involves both multiplication and addition. Napier's most striking result comes a few pages later:

'*Of the five parts of a spherical triangle, given the three intermediate, to find the two extremes by a single operation. Or otherwise, given the base and adjacent angles, to find the two sides.*

'Of the angles at the base write down the sum, half-sum, difference and half-difference along with their logarithms.

'Add together the logarithm of the half-sum, the logarithm of the difference and the logarithm of the tangent of half the base; subtract the logarithm of the sum and the logarithm of the half-difference, and you will have the first found.

'Then add the logarithm of the half-difference and the logarithm of the tangent of half the base, subtract the logarithm of the half-sum, and you will have the second found.

'Look for the first and second found among the logarithms of tangents, since they are such, then add their arcs and you will have the greater side; again subtract the less arc from the greater, and you will have the less side.'‖

This means, in our notation:

$$\log\sin\frac{A+B}{2}+\log\sin(A-B)+\log\tan\frac{c}{2}-\log\sin(A+B)-\log\sin\frac{A-B}{2}$$
$$=\log\tan\frac{a+b}{2},$$

$$\log\sin\frac{A-B}{2}+\log\tan\frac{c}{2}-\log\tan\frac{A+B}{2}=\log\tan\frac{a-b}{2}.$$

† q.v., vol. ii, pp. 14–16.   ‡ q.v., vol. v.   § Napier[1], p. 68.   ‖ Ibid., p. 63.

Removing the logarithms:

$$\frac{\sin\frac{1}{2}(A+B)\sin(A-B)\tan\frac{1}{2}c}{\sin(A+B)\sin\frac{1}{2}(A-B)} = \tan\frac{1}{2}(a+b).$$

Writing $\sin(A+B)$ and $\sin(A-B)$ in terms of the half-angles:

$$\frac{\cos\frac{1}{2}(A-B)}{\cos\frac{1}{2}(A+B)}\tan\frac{1}{2}c = \tan\frac{1}{2}(a+b). \tag{8}$$

In the same way we get from his 'second part':

$$\frac{\sin\frac{1}{2}(A-B)}{\sin\frac{1}{2}(A+B)}\tan\frac{1}{2}c = \tan\frac{1}{2}(a-b). \tag{9}$$

These are known as Napier's 'analogies'. He doubtless found them in hunting for something to correspond to Vieta's Law of tangents, given on p. 80. Analogous to these are the formulae found by Briggs:[†]

$$\left.\begin{aligned}
\frac{\sin\frac{1}{2}(a-b)}{\sin\frac{1}{2}(a+b)}\tan\frac{1}{2}C &= \tan\frac{1}{2}(A-B)\,; \\[2mm]
\frac{\cos\frac{1}{2}(a-b)}{\cos\frac{1}{2}(a+b)}\tan\frac{1}{2}C &= \tan\frac{1}{2}(A+B).
\end{aligned}\right\} \tag{10}$$

Napier's trigonometric work is directed towards a useful end, and the result is useful. I agree with the laudatory opinion of von Braunmühl:

'Überblicken wir Neper's Leistungen im Gebiete der Trigonometrie, so müssen wir zugestehen, dass durch seine Erfindung der Logarithmen diese Wissenschaft in ganze neue Bahnen geleitet wurde, und dass von ihm selbst schon die Richtung angegeben worden ist, nach welcher die Umgestaltung der bisher in Gebrauch befindlichen Sätze stattzufinden hatte, und das neue Instrument in fruchtbringender Weise zu verwerthen. Aber auch sein zweites Verdienst ist nicht gering anzuschlagen, dass es ihm zum erstenmal geglückt ist, die verwirrende der Sätze, die bisher zur Behandlung der rechtwinkeligen Kugeldreiecke dienten, durch eine klar und kurzgefasste Regel zu ersetzen.'[‡]

§ 3. Napier's rods

Napier's practical spirit, when questions of a theoretical sort did not come in, led him continually to look for labour-saving devices which would simplify calculation. Yet he himself can scarcely have appreciated the value of his logarithms, otherwise he would not have troubled himself, just about the same time, to develop certain other devices which were highly esteemed at the period, but which appear to us to-day as rather trivial. One of these was the use of strips of metal or bone to aid in multiplication and division. The description is found in Napier[4].[§]

† Napier[1], p. 80.   ‡ von Braunmühl (q.v.), II, 17.   § pp. 6 ff.

Let us imagine a long strip, marked at the top with one of the first nine integers, and in the squares below the product of this integer and each of the first nine. When the product involves two integers, these are separated by a diagonal line. Suppose, for instance, we wish to multiply 257 by 36. We place next to one another the rods for 2, 5, and 7, and carry through the multiplication, first by 6 and then by 3.

Napier begins on the line 6. If we multiply into 7 we get 42. Write down the 2, but carry the 4 diagonally downwards adding it to the 0; carry the 3 diagonally downwards and add to the 2, getting 5; write down the 1, thus getting 1542.

We multiply by 3 in exactly the same way, finally putting

$$
\begin{array}{r}
1542 \\
771 \\
\hline
9252
\end{array}
$$

Division is managed in somewhat the same way but is naturally more complicated. Suppose we wish to divide 354,526 by 257. We use the same rods for finding the partial products, guessing each time at the next figure of the quotient. Napier's work looks like this:

$$
\begin{array}{lll}
243 & & \\
204 & & \underline{123} \\
97 & & 257 \\
354526 & \quad 1379 & \\
257 & & \\
771 & & \\
1799 & & \\
2313 & &
\end{array}
$$

I should like to point out that this is essentially the usual process of long division except that partial products are written below and remainders above.

It is remarkable that Napier devised more complicated rods for extracting square and even cube roots. For finding a square root we have what he calls a 'lamina', giving in three columns the numbers x^2, $2x$, x. Here is how he finds the square root of 117716237694:[†]

$$90$$
$$54895$$
$$67$$
$$21$$
$$2$$
$$11.77.16.23.76.94$$
$$3\quad 4\quad 3\quad 0\quad 9\quad 8$$
$$9$$
$$256$$
$$2049$$
$$617481$$
$$5489504$$

Here is the *modus operandi*. The nearest root to 11 is 3, whose square is 9, written below; the difference is 2, written above. Take the rod marked 6. Find by the aid of it and the lamina the integer whose square plus its double comes nearest to 277 when multiplied by 30. This integer is 4, and the square and double add to 256, leaving a remainder 21, both set down in the proper places. We double 4, getting 8, take the rods for 6 and 8 and the lamina, and find the integer whose square and double multiplied by 340 come nearest to 2116. This integer is 3; we keep on in this way to the end. There is a similar but more elaborate system for cube roots.

Napier's inventive genius does not stop here. He describes a box, which he calls a 'Pyxis', in Napier[4].[‡] This does not seem to have attained much popularity. There is, indeed, not much to be said in favour of his *Rhabdiologia* as a contribution to mathematical theory: we can but wonder at the great esteem in which it was held by his contemporaries.

§ 4. *De arte logistica*

This is Napier's most elaborate mathematical work. It was put together from various fragments, some incomplete, which were left behind at his death. It was published in handsome style by Mark Napier, who remarks:†

'It would appear however, that his Algebra, as far as orderly set down, is an earlier production than either his Arithmetic or the fragment of Geometrical Logistic. . . . Napier adopts in his Algebra the nomination and notation which had been introduced before his time, whereas in the Arithmetic and Geometrical Logistic he adopts and expands a peculiar numeration of his own, applicable to the Arithmetic of Surds whereby he proposes to supersede that with which he operates in his treatise on Algebra. There can be little doubt, therefore, that his Algebra is a work of still earlier date than the other books, and these, as has been seen here, were prior in date to his conception of Logarithms which was some time before 1594.'

But, after all, what is logistic? In modern times it means the science of moving military forces, but Napier says: 'Logistica est ars bene computandi', the art of computing well. He goes ahead and defines the four fundamental operations, division being called *partitio*; it can be either perfect or imperfect. He early comes to the extraction of roots, and here, in the first part of the *Ars Logistica*, we find an ingenious notation of his own devising. We set the nine first integers in cells like this:

$$
\begin{array}{c|c|c}
1 & 2 & 3 \\
\hline
4 & 5 & 6 \\
\hline
7 & 8 & 9
\end{array}
$$

The boundaries of the cells are the radical signs for the roots whose indices are the numbers in the cells. Thus the square root of A is written $\sqcup A$, while its cube root is $\llcorner A$. This number cannot always be found, i.e. it is not always an integer. In chapter vi of Book I of Napier[3] we first come to grips with negative numbers, 'De quantitatibus abundantis et defectibus'. The signs $+$ and $-$ are introduced, and the rules for calculating with them rightly given. There is nothing essentially new here, negative numbers were in current use. He notes that an even root of an abundant number may be either abundant or defective. Book I ends with fractions. In Book II we come to closer grips with the actual labour of computation, first with integers. The first real difficulty comes with division. This is done by the old 'scratch and galley method', still in use in his time,‡ although, as we saw on p. 83, he was familiar with what is essentially our present practice. He then returns to radicals which exercise him greatly. The difference between the same two powers of two integers is called a *supplement*; thus 44 is a supplement, as it is the difference between the squares of 10 and 12. He undertakes to prepare a table of supplements, and this

---

† Napier[5], 1.   ‡ Smith (q.v.), vol. ii, pp. 136 ff.

amounts to writing the differences between a^n and $(a+b)^n$. This appears in the form of a triangular array:

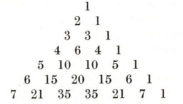

$$
\begin{array}{ccccccc}
1 \\
2 & 1 \\
3 & 3 & 1 \\
4 & 6 & 4 & 1 \\
5 & 10 & 10 & 5 & 1 \\
6 & 15 & 20 & 15 & 6 & 1 \\
7 & 21 & 35 & 35 & 21 & 7 & 1 \\
\end{array}
$$

This is usually called Pascal's triangle, after the writer whose use of it we shall discuss in the next chapter. We see at once that the numbers in any horizontal row give the coefficients in the binomial expansion. But we saw on p. 20 that there is reason to believe that Omar Khayyám may have known of this in the eleventh century: the Arabs knew it in the twelfth and the Chinese not only knew the law, but expressed it in triangular form in the fourteenth.[†] Napier probably knew nothing of all these people, but he received a good part of his education on the Continent, and may well have been familiar with the work of Stifel. Now Stifel shows a slightly modified form of this triangle,[‡] he knew the expansion up to $(a+b)^{17}$. Napier's placing this figure in his work has not the historical significance that some commentators seem to find. In approximating to a square root he uses what is essentially the inequality

$$
a+\frac{b}{2a+1} < \sqrt{(a^2+b)} < a+\frac{b}{2a}.
$$

This, however, is an old Arabic rule known to Leonardo of Pisa and Tartaglia.[§] Napier also gives an approximation to a cube root which is erroneous.[||] There follows a discussion of fractions ending with 'De fractionibus physicis'. These are what we call 'denominate numbers'; the discussion is very short.

The third book deals with geometrical logistics: 'Geometrica ergo dicitur logistica quantitatum concretarum per numeros concretos. Concretus dicitur omnis numerus quatenus quantitatem concretam et continuam referat.' This means, in modern phrase, that he is going to discuss the continuum; continuous quantities are those which can be represented by the lengths of line segments. He reintroduces his notation for roots, and is especially interested in monomials. 'Unde sequitur quod uninomium vel est numerus unicus simplex, vel unici numeri simplex radix aliqua.'[††] He gives as examples ⊔10, ⌊12, ⊐26. He then points out that roots can be positive or negative or both, as the

† Tropfke (q.v.), 2nd ed., vol. vi, p. 35.   .‡ Stifel (q.v.), p. 44.
§ Cajori², p. 150. There occurs a mistake of sign in this.
|| Stegell in Napier², p. 152.   †† Napier⁵, p. 85.

even roots of positive numbers; as for the even roots of negative numbers, 'Quaedam tandem nec sunt abundantia nec defectiva, quae nugacia vocamus. Hujus arcani magni algebraici fundamentum superius Lib. 1 cap. 6 jecimus (quod quamvis a nemine quod sciam revelatum sit) quantum tamen emolumenti adfert huic arti, et caeteris mathematicis, postea patebit.'[†]

This shows clearly that he believes he is revealing a great secret, and that no one before him had considered anything like his *nugacia*. I see no reason to doubt his good faith in the matter, and the same view has been expressed by others, as Mark Napier:

'There can be no doubt that by nugacia Napier means the impossible quantity, and that he was the very first to conceive the idea and propose its use in the arithmetic of surds and in the theory of equations. . . . The great emolument which Napier expected to bestow on Mathematics by this ghost of a quantity can only be understood by profound mathematicians.'[‡]

The same erroneous view is held by Stegel: 'But however this may be, there is no doubt that Napier's reference to imaginaries is the first on record.'[§] I am afraid that there is a great deal of doubt. The idea was certainly present to the mind of Cardan: 'Secundum genus positionis falsae, est per radicem $\bar{m}$. Et dabo exemplum, si quis discat, divide 10 in duas partes, ex quarum unius in reliquam ducto producatur 30 aut 40.'[||]

He puts through the calculation and finds $5\bar{p}R_x\ \bar{m}15$, $5\bar{m}R_x\ \bar{m}15$, that is to say, $5+\sqrt{-15}$, $5-\sqrt{-15}$.

Let us now inquire as to what Napier makes of his great secret. We read, p. 86 of Napier[5], 'In nugacibus sumponere cavendum est ne copula munitionis interponendum praeponitur'. This means that we must not confuse $\sqcup-9$ with $-\sqcup9$; beyond this he has naught to say.

The last part in position, but the earliest in composition, of the *Ars Logistica* deals with Algebra: 'Algebra scientia est de questionibus quanti et quoti solvendi tractans.' This means that Algebra is the science of finding numbers from facts known about them. His first worry is about monomial surds. Two surds are said to be commensurable if they are of the same order and if the quotient of their radicands is a perfect power of that order. He gives as an example $\sqcup12$, $\sqcup3$. The familiar rules follow, but sometimes he falls into error: 'Ut sit $10-\sqcup3$ dividendum per $6+\sqcup2$ quod non aliter fit quam interlineali divisione

$$\frac{10-\sqcup3}{6+\sqcup2}.$$

In a small marginal note is written: 'Haec sunt emendanda, nam per

† Ibid., p. 85.   ‡ Ibid., p. lxxxii.   § Napier[2], p. 155.
|| Cardan (q.v.), t. iv, p. 287.

$6+\sqcup2$ fieri potest divisio ut per omne binomium ex fine praecedentis capitis.' The note is not perfectly correct either, for there is no hint in the preceding chapter how we should have to rationalize the denominator if it involved a cube root instead of a square root.†

Book II of the Algebra is longer, and harder to understand. He uses certain characters rather indiscriminately for his unknowns. His favourite symbol is R_x. I am not sure whether this stands for Res or Radix, as in Cardan. The coefficient is sometimes prefixed. He indicates powers by certain symbols preceded by $\sqrt{}$. These symbols are called *Numeri ordinis*, and he knows that when we multiply two powers we should add the symbols.‡ In other words, he was in possession of the essentials of exponential notation, but his symbolism was sadly inefficient, and the same facts had been previously discovered by others.§

I cannot, on the whole, bestow high praise on the *Ars Logistica*, either as a contribution to pure mathematics or as an aid to calculation. But I do not wish to close my appreciation of this great man with a disparaging note, for a great man he certainly was, even though the statement that he was the greatest son of Scotland may be excessive. No man has done more to lighten the labours of those of his fellow men who were condemned to the work of calculation. By his patience and ingenuity in using the decimal point, in extending the use of decimal fractions, and above all in inventing, perfecting, and calculating logarithms, he rendered a service of incalculable value to all succeeding generations.

† Napier[5], p. 106.   ‡ Ibid., p. 127.
§ Cf. Smith in Napier[2], pp. 81 ff., and Clavius (q.v.), p. 16.

CHAPTER VII

BLAISE PASCAL

§ 1. Pascal's theorem

EVERY writer who deals extensively with the history of mathematics pays considerable attention to the famous theologian, philosopher, and master of French style whose name appears above. He was born in 1623, and died in 1662. His originality and ability were remarkable; had he confined his attention to mathematics he might have enriched the subject with many remarkable discoveries. But after his early youth he devoted most of his small measure of strength to theological questions, and his contributions to philosophical thought, and to fixing French prose outweigh in importance what he has done for pure science. He is not out of place when classed as a great amateur.†

There are various romantic legends concerning Pascal's life which we need only mention briefly. His father, Étienne Pascal, was a man of good mathematical ability. He was much worried by his son's delicate constitution and devotion to his books, so he prohibited the boy, not only from studying geometry, but even from thinking about it, merely telling him it was the science of drawing figures correctly. The prohibition was enough to set young Blaise to thinking intently on the subject, and the story has it that by himself he worked out many of the theorems of elementary geometry, including that which gives the sum of the angles of a triangle. When this precocity was discovered, the prohibition was removed, and the young Pascal made surprising progress with his studies. At the age of fourteen he attended scientific meetings with such men as Mydorge, Mersenne, Roberval, and others; at sixteen he composed a work on the conic sections which must have been altogether remarkable. We have the authority of Mersenne for the statement that from his theorem of the hexagon, or *Hexagramma mysticum*, he drew over four hundred corollaries.‡ We are, I think, safe in classing this as a fable, but it must have been a remarkable piece of work, and it is completely lost! All that we have now is a short fragment, composed in 1640, called 'Essai sur les coniques'. In this we find the theorem, 'dont l'inventeur est M. Desargues Lyonnois, un des grands esprits de ce temps', that a transversal will meet a conic, and the pairs of opposite sides of an inscribed quadrangle, in pairs of points of an involution. The mere fact that Pascal appreciated the work of that strange man,

---

† For a detailed list of Pascal's mathematical writings see Marie (q.v.), vol. lv, p. 185 f.
‡ Cantor[1], vol. ii, pp. 621 ff.

Girard Desargues, is a proof of his capacity. We find here also his
hexagramma mysticum, or inscribed hexagon, together with:

Pascal's Theorem] *If the vertices of a hexagon lie on a conic, the intersections of the pairs of opposite sides lie on a straight line.*

The converse of this theorem is also true, and the theorem holds even
when the conic is degenerate.

How did Pascal prove his theorem? We have only one hint. He
stated it originally only in the case of a circle, and then remarked that
as it is projectively invariant it holds for the general conic which, by
definition, is the projection of a circle. It is therefore probable that he
proved it for the circle, using some property, like the equality of all
angles inscribed in the same circular arc, which is peculiar to that curve.
Proofs based on this have subsequently been developed; they are among
the most difficult proofs of the theorem. We can, without difficulty,
reconstruct proofs that would have been quite accessible to Pascal.
That given by Brianchon (q.v.) is very much in the manner of Desargues,
whom Pascal admired so much, and the latter might well have found it,
but then he would not have stated it first for the case of a circle. It is
to be noted also that he states his theorem in very clumsy shape, showing that a pair of opposite sides of an inscribed hexagon and the line
connecting the intersections of the other two pairs of opposite sides
'sont de mesme ordre'. Now he begins the present essay with the
words 'Quand plusieurs lignes droictes concourent au mesme point ou
sont toutes parallèles entre elles, ces lignes sont dictes de mesme ordre
ou de mesme ordonnance', a clumsy wording taken from Desargues.†

§ 2. The logic of mathematics

In the present work I shall confine myself strictly to the study of
Pascal's work as a pure mathematician, omitting all reference to his
study of the pressure of the air, even of his calculating machine. I may
not, however, omit all reference to his discussion of the philosophy of
mathematics, which is included in his *Pensées*, in an essay 'De la
démonstration géométrique'.‡ Here he undertakes a careful analysis of
the essential quality of mathematical reasoning. Pascal was nothing if
not a casuist: argument was meat and drink to him; he was interested
in the principles of demonstration, and he found them in geometry.
The basis of accurate discussion must be accurate definition. He begins
by refuting those who say that definition is unnecessary, and those who
claim that everything must be defined. It is perfectly clear that in any
definition the meaning of some terms must be assumed as known at the
outset. In modern abstract mathematics we cut the number of these
down as low as possible, but we assume that we know what we mean

† Pascal (q.v.), vol. i, p. 251.   ‡ Ibid., vol. ix, pp. 271 ff.

by such purely logical terms as 'exist', 'one', 'none', 'belong to', etc. Pascal is more liberal. He says:† 'Elle ne définit aucune de ces choses espace, temps, mouvement, nombre, égalité ni les semblables qui sont en grand nombre, par-ce-que ces termes là désignent si naturellement les choses qu'elles signifient, à ceux qui entendent la langue, que l'éclair-cissement qu'on voudrait en faire apporterait plus d'obscurité que d'instruction.' He gives as an example the absurdity of one who would define light as a luminous movement of shining bodies. He takes a pot-shot at the classical difficulty of the infinite divisibility of space. He is especially caustic at the expense of one who maintains that space can be divided into two parts which are themselves indivisible. He is emotionally much moved by the thought of the infinitely large and the infinitely small.

The second part of the essay is called 'De l'art de persuader'. Here we get rules for definitions, axioms, and demonstrations. He finally gets down to this:‡

'Règles pour les définitions.

'I. N'entreprendre de définir aucune des choses tellement connues d'elles-mesmes, qu'on n'ait point de termes plus clairs pour les expliquer.

'II. N'omettre aucun des termes un peu obscurs ou équivoques sans définition.

'III. N'employer dans la définition des termes que des mots parfaitement connus ou déjà expliquéz.

'Règles pour les axiomes.

'I. N'omettre aucun des principes nécessaires sans avoir demandé si on l'accorde, quelque clair et évident qu'il puisse être.

'II. Ne demander en axiomes que des choses parfaitement évidentes d'elles-mesmes.

'Règles pour les démonstrations.

'I. N'entreprendre de démontrer aucune des choses qui sont tellement évidentes d'elles-mesmes, qu'on n'ait rien de plus clair pour les prouver.

'II. Prouver toutes les propositions un peu obscures, et n'employer à leur preuve que des axiomes très évidens, ou des propositions déjà accordées ou demonstrées.

'III. Substituer toujours mentalement les définitions à la place des définis, pour ne pas se tromper par l'équivoque des termes que les définitions ont restreints.'

This is not exactly the way that we should phrase things to-day. It took nearly three hundred years of mathematical thought to reach the modern abstract logical point of view. We do not begin our geometry by defining points, nor do we omit all definitions because it is perfectly clear what they are. We do not much care what they are, provided we may make certain purely logical assumptions about them. We are little interested in the truth of axioms, but much in their independence. Pascal was a long way from Hilbert and Russell. But I accept gladly Cantor's judgement: 'Der erste moderne Versuch einer Philosophie der Mathematik'.† The *Pensées* did not appear till 1669, but were probably written before Arnauld's *Logique de Port-Royal*, which contains many similar ideas, as we shall see in the next chapter, and which appeared in 1662. Doubtless the two writers shared their thoughts in these as in other matters.

§ 3. The arithmetical triangle

All who have written about Pascal as a mathematician have written about the arithmetical triangle, to which his name is usually attached. We saw on p. 86 that not only was this known to Stifel, but much earlier to the Chinese, but Pascal was so much interested in it that it is well to discuss it in connexion with his work. In Fig. 41 we reproduce Pascal's own drawing. We can describe it very succinctly as follows. Let us imagine that the positive X-axis has been run through the centres of the top row of unit squares, and the negative Y-axis through the centres of the left columns of unit squares. With this arrangement, at the point whose coordinates are $(r, r-n)$, $n \geqslant r$, we write down

$$\binom{n}{r} \equiv \frac{n!}{r!\,(n-r)!}. \tag{1}$$

Pascal does not describe it in this way. In the first 'rang parallèle' we write a series of units, in the second the natural series, whose differences are the members of the first series. In the third row we have the so-called 'triangular series' whose differences are the second series, in the fourth the 'pyramidal series' whose differences are the triangular one, and so on. These so-called 'figurate numbers' go back to the Greeks, as well as their names; they appear in the same connexion in Stifel.‡ The 'rangs perpendiculaires' will be built on the same principle. The number in any square is the sum of the numbers immediately above, and immediately to the left, which gives

$$\binom{n}{r} = \binom{n-1}{r-1} + \binom{n-1}{r}. \tag{2}$$

Pascal finds a number of other identities as

$$\binom{n}{r} \div \binom{n}{r-1} = \frac{n+1-r}{r}; \qquad \binom{n}{r} \div \binom{n-1}{r-1} = \frac{n}{r}. \tag{3}$$

The most striking feature of the triangle is that the numbers in any north-easterly running diagonal are binomial coefficients. Pascal him-

† Cantor[1], vol. ii, p. 682.   ‡ Euclid (q.v.), vol. ii, pp. 289 ff.

self points this out.† I do not think that we should for this reason set him down as the discoverer of the binomial theorem. I mentioned on p. 20 that there is good reason to believe that the credit should go to Omar Khayyám, and as Stifel used these same numbers to extract roots

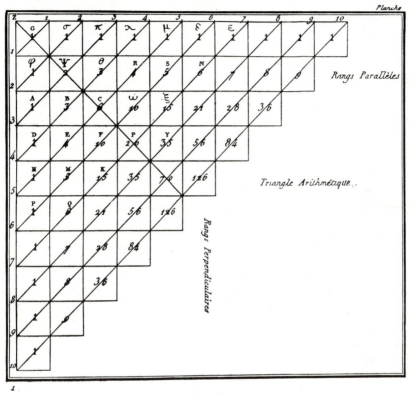

Fig. 41

of varying orders, it seems likely that he connected them with the binomial theorem.‡ Pascal calls the different rows 'ordres numériques', and as I said on the last page develops many identities from them. More interesting is his application to the theory of combinations. The number of groups of r objects taken from n objects is $\binom{n}{r}$. Here is Pascal's statement:

'Le nombre de quelque cellule que ce soit égale la multitude des combinaisons d'un nombre moindre de l'unité que l'exposant de son rang parallèle dans un nombre moins de l'unité que l'exposant de sa base.'§

† Pascal (q.v.), vol. iii, p. 499.
‡ Stifel (q.v.), sheet 44; also Cantor[1], vol. ii, p. 234.
§ Pascal (q.v.), vol. iii, p. 476.

The confusing phrase *moins de l'unité* is explained by the simplification I have made in the notation by a slight sliding of the axes. In any case the formula for the number of combinations of n things taken r at a time had been found long before.†

Much more original is the application of the arithmetical triangle to the calculus of probability. This branch of mathematics was coming rapidly to the fore in Pascal's time. He exchanged letters on the subject with Fermat, the Chevalier de Méré, and others. The particular application of the arithmetical triangle is to the problem of how the stakes should be divided when two players, of supposed equal skill, are forced to break off a game before it is finished. This problem was brought to Pascal by Chevalier de Méré, who confessed that he could not solve it in the general case. Pascal first takes up some special cases. Suppose that the first is within one point of winning, while the second needs two points. The first has a half-chance of winning the first turn, in which case he has won the whole, and a half-chance of losing the first turn, in which case they are equally placed. The chance of the first is then $\frac{1}{2}+\frac{1}{2}\cdot\frac{1}{2}=\frac{3}{4}$.

Suppose secondly that the first has one point only needed to win, the second three. If the first wins the first point he is finished, if he loses the situation of the preceding problem has arisen; we have then $\frac{1}{2}+\frac{1}{2}\cdot\frac{3}{4}=\frac{7}{8}$.

Or suppose the first needs two to win and the second three. If the first wins the first point the last situation has arisen, if he loses it, they are equally well off. The chance for the first is thus $\frac{1}{2}\cdot\frac{7}{8}+\frac{1}{2}\cdot\frac{1}{2}=\frac{11}{16}$.

Now for the general case. Pascal and Fermat both worked at this. The latter confessed that it was beyond him. Here is Pascal's solution, which I have shortened as his reasoning is rather tedious.‡

Let the first need m to win and the second n, and let us call the chance for the first $f(m, n)$. Let us write $m+n-1 = r$. Pascal assumes

$$f(m, n) = \frac{1}{2^r}\left[\binom{r}{0}+\binom{r}{1}+\dots+\binom{r}{n-1}\right], \tag{4}$$

$$\binom{r}{0}=\binom{r+1}{0}=1. \tag{5}$$

Now let us find $f(m+1, n)$. There is a half-chance that he will win the first time, in which case he will need only m to win, or he may lose the first time, in which case the adversary will only need $n-1$. Hence

$$f(m+1, n) = \tfrac{1}{2}[f(m, n)+f(m+1, n-1)].$$

† See editorial comment, Pascal (q.v.), vol. iii, p. 442.
‡ Pascal (q.v.), vol. iii, pp. 490 ff.

Substituting,

$$\frac{1}{2^{r+1}}\left[\binom{r}{0}+\binom{r}{1}+\cdots+\binom{r}{n-1}+\binom{r}{0}+\binom{r}{1}+\cdots+\binom{r}{n-2}\right].$$

Now, by (3),
$$\binom{r}{k}+\binom{r}{k-1}=\binom{r+1}{k}.$$

Hence we have

$$\frac{1}{2^{r+1}}\left[\binom{r+1}{0}+\binom{r+1}{1}+\cdots+\binom{r+1}{n-1}\right],$$

and this shows that formula (4) steps up. The same will hold when we go from n to $n+1$.

Cantor has pointed out that we have here a perfect example of mathematical induction, and expresses the belief that this is the first case on record.[†] This opinion is incorrect, as Vacca pointed out in a letter; the method had been used by Maurolycus in the sixteenth century. Now Pascal knew of Maurolycus, for he quotes 'Maurolic' in a letter to M. de Carcavi.[‡] This suggests that Pascal may have taken the method over bodily without mentioning it. Let us hope that such was not the case. For a further discussion see Bussey (q.v.).

§ 4. Centres of gravity

I will now return to Pascal's orders of numbers. In a letter to M. de Carcavi[§] he proceeds as follows. Suppose that we have a set of quantities $A, B, C,\ldots$. Then $A+B+C+\ldots$ is called their *simple sum*. We then take the sums

$$A+B+C+D+\ldots$$
$$B+C+D+\ldots$$
$$C+D+\ldots$$
$$D+\ldots.$$

These added give

$$A+2B+3C+4D+\ldots,$$

which he calls the *triangular sums*. Similarly we take

$$A+2B+3C+4D+\ldots$$
$$B+2C+3D+\ldots$$
$$C+2D+\ldots$$
$$D+\ldots,$$

giving their *pyramidal sum*

$$A+3B+6C+10D+\ldots.$$

[†] Cantor[1], p. 749.   [‡] Pascal (q.v.), vol. viii, p. 343.
[§] Ibid., pp. 337 ff.

The triangular sum is the more interesting, so I confine my attention to that. Let us imagine that these quantities are weights, strung at unit distances along a straight line. The moment about a point a unit's distance from the first point will be the triangular sum. The moment about any point will be that about any other point plus the moment about the former of the whole weight if concentrated at the latter. This is more concise than Pascal's statement which does not include the possibility of negative weights.

Let us now consider the moments of an infinite number of infinitesimally small quantities, as the moment of a curve about an axis. We must first say something further about such infinite sums. It must be borne in mind that Pascal was intimate with Roberval who, following Cavalieri, was deeply steeped in the idea of indivisibles. The mathematical world was at this time feeling its way towards the infinitesimal calculus. Pascal's ideas were essentially sound. One might judge at first reading that Pascal believed that an area was the actual sum of an infinite number of line segments, or lines as he would have said, but this was not exactly the case. He means an infinite sum of infinitesimal rectangles. He does not mention the 'limit of a sum'—mathematical thought had not reached that point—but he was very near to grasping the idea. Moreover he draws very important distinctions which show that he did not add sums in the straightforward manner. Suppose that we divide that portion of the X-axis which lies under a continuous curve into an infinite number of equal lengths, that is to say, an ever increasingly large number of such lengths. The perpendiculars to these lengths are what he calls 'ordinates', and when he speaks of their sum he means what we should write $\int y \, dx$. But suppose that instead of dividing up the base into equal segments we divide the curve itself, and drop perpendiculars on the axis from these points. These perpendiculars are called 'sines to the base', and when he speaks of a sum here he means $\int y \, ds$. The fact that there are the same number of divisions in the two cases is immaterial. We shall see later that he will speak of the 'abscissa of centre of gravity of ordinates to base', meaning $\bar{x}$ where

$$\int (x-\bar{x})y \, ds = 0.$$

He is interested in both simple and triangular sums, the latter, of course, moments. As an example of the way he goes to work, let me show how he determines the centre of gravity of an arch of the cycloid.† This clearly lies on the middle vertical, so we have merely to determine its ordinate; we may confine our attention to one half of the curve. Let us find (Fig. 42) the moment about a tangent at the top of the arch. We have:

$$\text{Moment} = 2 \int CZ \, ds.$$

† Pascal (q.v.), vol. viii, p. 359.

By a familiar property of the cycloid:

$$\text{Arc } CY = 2CM; \qquad CM^2 = CZ.CF = OZ^2;$$

where O lies on a parabola with axis CF.

$$\text{Moment} = 2 \int CZ \, dOZ = 2 \int x \, dy.$$

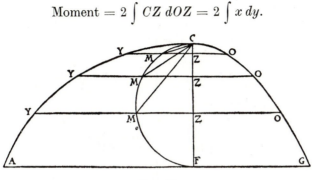

FIG. 42

This is the area of the parabolic half-segment $= \frac{4}{3}CF^2$.

But $\qquad\qquad\qquad\qquad\qquad$ Arc $CA = 2CF$.

Hence $\qquad\qquad\qquad\qquad\qquad$ $\bar{y} = \frac{2}{3}CF$.

§ 5. Integration

I shall return later to Pascal's study of the cycloid. First, however, I wish to take up what seems to me altogether the most interesting and significant part of his mathematical work, his study of definite integrals. His general scheme is to make some such integral correspond to the area or volume of some simpler figure, and then determine the latter by a change of variable. His fundamental figure is called a 'triline' or 'triligne'. This is an area bounded by a smooth arc and two mutually perpendicular line segments called the axis and the base. If the curve is $x = f(y)$ the moment about the X-axis of the area is $\int xy \, dy$.

Now through each point of the triline we erect a perpendicular to the plane. Through the base we pass a plane making an angle of 45° with the plane of the triline. The volume contained between the two planes and the cylindrical surface which he calls a notch ('onglet') will be $\int xy \, dy$. If we can find this volume by some other integration, we can find the moment of the triline.

Pascal made use of something even more complicated than the triline and its notch. Let us suppose (Fig. 43) that each abscissa is extended beyond the Y-axis until it meets a curve, called an 'adjoint' of the triline, in a point with the generic denomination I. The figure AIK is then turned about AB as an axis through an angle of 90°, so that it is

in a plane perpendicular to the plane of the triline. Through each point
of the arc BC and of the line-segment AC a perpendicular to the plane
of the triline is drawn, thus making a cylindrical surface. We cut this
surface by lines through all the points to AIK in its new position

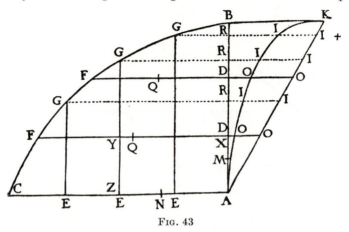

Fig. 43

parallel to AC, and seek the volume of this solid. We can find this in
two different ways.

First we can cut it by planes perpendicular to AC. The area in such
a plane is the curvilinear triangle ARI. On the other hand, we might
take planes perpendicular to AB. In this case the element of area is
the rectangle $GR.RI$. We thus get

$$\int_0^{AB} (ARI)\,dx = \int_0^{AB} x.RI\,dy. \tag{6}$$

Suppose, however, we do not wish for the whole volume, but so much
as lies below a chosen line GR, where $y = y_1$. Let z be the area of the
curvilinear triangle ARI. This, like x, is a function of y, and $IR.dy = dz$.
Our volume is then $\int_0^{y_1} x\,dz$.

But we might proceed otherwise. The volume is divided into two
parts. One is $RGAE$, a horizontal column standing on $ARI = z_1$, height,
section, etc. The other part is CGE. Here if we take the negative
direction as positive for X and upwards as positive for Y, as y increases
dx is negative. We thus get the final equation

$$\int_0^{y_1} x\,dz = x_1 z_1 - \int_0^{y_1} z\,dx. \tag{7}$$

This, of course, is the usual formula for integration by parts.† Pascal

† Cantor[1], vol. i, p. 838.

did not appreciate its full significance, neither did he understand the full significance of integration anyway. But in spite of his complicated method, he deserves full credit for discovering this fundamental formula.

Pascal makes applications to cases where the adjoint is a specified curve. Let it be the straight line

$$y = x'$$

$$\tfrac{1}{2} \int_0^{AB} y^2 \, dx = \int_0^{AB} xy \, dy. \tag{8}$$

Or suppose it is the parabola

$$y^2 = AB.x':$$

$$ARI = \frac{1}{3AB} y^3,$$

$$\int x.RI \, dy = \frac{1}{AB} \int xy^2 \, dy,$$

$$\tfrac{1}{3} \int_0^{AB} y^3 \, dx = \int_0^{AC} xy^2 \, dy. \tag{9}$$

Pascal finds sums of sines by a simple device. We start with the curve $y = f(x)$ and take also

$$x' = s; \qquad y' = y;$$

$$\int s \, dy = \int y \, ds; \qquad \int xy \, dy = \tfrac{1}{2} \int y^2 \, ds. \tag{10}$$

§ 6. The cycloid

As an introduction to Pascal's study of this curve, we study a half-segment of a circle of radius r:

$$\frac{dx}{ds} = \frac{y}{r}; \qquad r \, dx = y \, ds; \qquad r \, dy = -x \, ds;$$

$$\int_0^{s_1} y \, ds = r \int_0^{x_1} dx = rx_1; \qquad \int_0^{s_1} y^2 \, ds = r \int_0^{x_1} y \, dx = r.\text{area } ABRI; \tag{11}$$

$$\int_0^{s_1} y^3 \, ds = r \int_0^{x_1} y^2 \, dx = r \int_0^{x_1} (r^2 - x^2) \, dx = r^3 x_1 - \frac{rx_1^3}{3}. \tag{12}$$

Let $\bar{x}$ be the abscissa of the centre of gravity of the sines

$$0 = \int_0^{s_1} (x - \bar{x}) y \, ds = r \int_0^{x_1} (x - \bar{x}) \, dx; \qquad \bar{x} = \frac{x_1}{2}.$$

Now take a half-segment BSR:

$$x' = x; \qquad y' = y-a;$$

$$\int_0^{s_1} y' \, ds = rx_1 - as_1;$$ (13)

$$\int y'^2 \, ds = r \,.\, \text{area } ABRI - 2arx_1 + a^2 s_1;$$ (14)

$$\int xy' \, ds = \frac{rx_1^2}{2} - a \int_0^{s_1} x \, ds = \frac{rx_1^2}{2} + ar(r-a).$$ (15)

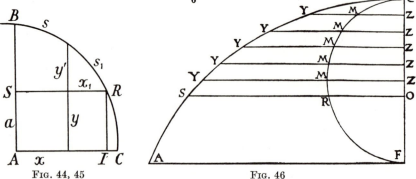

FIG. 44, 45   FIG. 46

Now for the cycloid. This curve, called by various names, had been studied not a little by his contemporaries, and men with whom he held scientific correspondence, especially Roberval. It was therefore natural that he should apply his methods of integration, which I have described, to this curve. He thereupon issued a challenge to contemporary mathematicians to solve, within a specified time, certain problems which he proposed. The following were given out in June 1658.

Given the triline whose curved part is an arch of a cycloid measured from the summit, the base is not necessarily the base of the cycloid. To find the content, and centre of gravity of the triline, of the solids obtained by rotating it about the base and axis, and of the zones so generated.

Neither of the contestants who sent in answers fulfilled the conditions imposed, as Pascal explains at great length.† Here are some of the integrals involved:

$$\int x \, dy, \quad \int xy \, dx, \quad \int x^2 y \, dy, \quad \int xy \, ds.$$

The central fact of which Pascal makes use is that

$$YZ = MZ + \text{arc } MC,$$

$$x = x' + s \qquad y' = y,$$

$$\int x \, dy = \int x' \, dy + \int s \, dy = \int x' \, dy + \int y \, ds$$

$$= ORC + r \,.\, OR - FO \,.\, \text{arc } CR.$$

† Pascal (q.v.), vol. ix, pp. 116 ff.

Other integrals are calculated by similar devices, all turning on the calculations for circular arcs we have shown. In all of this the methods are more interesting than the results. I have the uncomfortable feeling that Pascal picked his challenge problems to show his skill in a very limited field rather than to advance mathematical science. Perhaps this happened in other cases where one mathematician challenged another. One's final impression is not inspiring.

§ 7. Short studies

Pascal made many other studies in pure mathematics, and carried the subjects I have touched much farther. I will, however, only give two other examples of his work, one geometrical, one algebraic, or rather arithmetical. He makes an interesting comparison of the spiral of Archimedes with the parabola. We have such a spiral and parabola which begin at the same point and touch there. In modern notation we write them

$$x^2 = 2my; \qquad r = m\phi.$$

The points where $x = r$ are said to be corresponding.

$$\frac{r\,d\phi}{dr} = \frac{r}{m} = \frac{x}{m} = \frac{dy}{dx}.$$

We see thus that at corresponding points the tangent to the parabola makes the same angle with the tangent at its vertex that the tangent to the spiral makes with the radius vector. We have, moreover,

$$ds^2 = dr^2 + r^2\,d\phi^2 = dr^2\left(1 + \frac{r^2}{m^2}\right) = dx^2 + dy^2.$$

Corresponding arcs on the two curves have the same lengths.

The arithmetical note has to do with divisibility. When will the number

$$a_0 + 10a_1 + 100a_2 + 1000a_3 + \dots$$

be divisible by x which I suppose < 10.

Let

$$10 = q_1 x + r_1$$
$$10r_1 = q_2 x + r_2$$
$$10r_2 = q_3 x + r_3$$

$$\cdot \quad \cdot \quad \cdot \quad \cdot \quad \cdot$$

Then

$$100 = (10q_1 + q_2)x + r_2$$
$$1000 = (100q_1 + 10q_2 + q_3)x + r_3$$

$$\cdot \quad \cdot \quad \cdot \quad \cdot \quad \cdot \quad \cdot \quad \cdot$$

$$10^k = (10^{k-1}q_1 + 10^{k-2}q_2 + \dots + q_k)x + r_k.$$

The given number is divisible by x if

$$a_0 + a_1 r_1 + a_2 r_2 + \ldots + a_k r_k \equiv 0 \ (\text{mod}\, x).$$

The condition is theoretically correct; in practice I cannot see that it shortens in any way the problem of divisibility.

What should be our final judgement of Pascal as a pure mathematician? That he was unusually keen and original there can be no possible doubt. He came near to making great discoveries, but fell a little short of doing so. Very little that he wrote caused a noteworthy advance in mathematical science. He has a perfect claim to priority in the discovery of the theorem which bears his name, and a very pretty theorem it is, but not epoch-making. We have no clue as to what may have been included in the fabulous number of corollaries which he is said to have deduced from it; that there were anything like four hundred of them I very much doubt. He was one of the first to make use of mathematical induction, but not the actual discoverer, and it seems quite possible that he took it bodily without credit from Maurolicus. He really discovered integration by parts, but failed to see the real importance of it. I think that brilliant amateur in mathematics is the best description of this remarkable man.

CHAPTER VIII

ANTOINE ARNAULD

§ 1. General introduction

IT is really noteworthy that the small fraternity of the Port Royal produced two writers of uncommon mentality who had more than a passing interest in mathematics. One of these was Blaise Pascal, whom we discussed in the last chapter, and who stood very close to the company without being technically a member of it, the other was his intimate associate Antoine Arnauld, 'le grand Arnauld' as his countrymen frequently called him. His intimacy with Pascal was so close that in some places it is difficult to distinguish between them, a difficulty which has arisen more than once in the long history of mathematics. Pascal had original, brilliantly original, ideas, and made important contributions towards the advancement of mathematical science. Arnauld corresponded with Descartes and Leibniz, and doubtless appreciated the importance of their mathematical work, though his letters seem to have been more occupied with philosophical than with mathematical questions; nevertheless he can be classed as one who contributed to the advancement of mathematics, even though his contributions were of pedagogical rather than of scientific importance. A very complete account of his work will be found in Bopp[1].

We may divide Arnauld's mathematical writings into two parts, Arnauld[1] and Arnauld[2]. We must always bear in mind that he was a theologian, a very combative theologian, whose main interest in life was theological controversy. He openly proclaimed that the one source of certitude in this world is authority, and when authority speaks in clear tones there is no room for argument. But he believed also that man is a rational being, that truth can be learnt by observation and reason, and that there can be no higher exercise of our intellectual powers than that of reaching new truth by sound reasoning. It is therefore not surprising that such a man should write a very influential work on logic.

§ 2. *La Logique de Port-Royal*

This work, which bears the sub-title of 'L'Art de Penser', is not a complete treatise on the subject, for, as Jourdain points out in the preface to Arnauld[1],† there is no mention of inductive reasoning, but this commentator also points out, very justly: 'On ne peut pas apporter dans l'exposition des arides préceptes de la logique, plus d'ordre,

† p. xvii.

d'élégance et de clarté qu'Arnauld, un discernement plus habile de ce qu'il faut dire parce qu'il est nécessaire, et ce qu'il faut taire, parce qu'il est superflu.' Now mathematics, which the French frequently call *géométrie*, is the shining example of what can be accomplished by deductive reasoning. Much of the work on logic is, consequently, devoted to a criticism of the foundations and teaching of this subject.

Arnauld's discussion of the philosophy of mathematics appears first in the fourth part of Arnauld[1] which deals with 'La méthode'. He insists that we have to acknowledge the existence of certain things which are in themselves inconceivable, because the denial leads to error. He argues strongly for the existence of infinite divisibility, hard as that may be to imagine. No integer, which is a perfect square, is double another which is also a perfect square, yet we can construct a square which has double the area of another square. But if there existed an indivisible small unit of area, the number of units in one would be double the number of units in the other. It is also hard to imagine something which is at once finite and infinite, but here is an example, which Arnauld stigmatizes as *assez grossière*.† We take a rectangle which is half of a square, lay one-half of itself next, end to end, then one-half of the half, and so on. The limit of the length is infinite, but the limit of the area is that of the square. Such examples are ingenious, even if not very profound.

Arnauld recognizes in the second chapter two methods which he calls analysis and synthesis. One would expect to find references to Plato, but there are none. He has, however, the same idea in mind, he merely illustrates it differently. Suppose we wish to prove that a certain man is descended from Saint Louis. One method would be to show that he is the child of A_1 and B_1, who are the children of A_{12}, A_{11} and of B_{12}, B_{11}, and so on back until one arrives at the sainted king. But one might also begin with the King of France, give the list of his children, of their children, and so on down until one finally came to the individual in question. He also mentions the method of *reductio ad absurdum* as an example of analysis.

In chapter iii we have certain rules for mathematical demonstration which appear in greater detail in chapter xi. I will give them at length:

'Deux règles touchant les définitions

'1) Ne laisser aucun des termes un peu obscurs ou équivoques sans les définir.

'2) N'employer dans les définitions que des termes parfaitement connus ou déjà expliqués.

'Deux règles pour les axiomes

'3) Ne demander en axiomes que des choses parfaitement évidentes.

† Arnauld[1], p. 272.

'4) Recevoir pour évident tout ce qui n'a besoin que d'un peu d'attention pour être reconnu véritable.

'Deux règles pour les démonstrations

'5) Prouver toutes les propositions un peu obscures, en n'employant pour leur preuve que les définitions qui auront précédé et les axiomes qui auront esté accordés, ou les propositions qui auront esté démontrées.

'6) N'abuser jamais de l'équivoque des termes en manquant de substituer mentalement les définitions qui les restreignent et les expliquent.

'Deux règles pour la méthode

'7) Traiter les choses, autant qu'il se peut, dans leur ordre naturel, en commençant par les plus générales et les plus simples, et expliquant tout ce qui appartient à la nature du genre, avant que de passer aux espèces particulières.

'8) Diviser, autant qu'il se peut, chaque genre en toutes ses espèces, chaque tout en toutes ses parties, chaque difficulté en tous ses cas.'

There are various comments that occur at once. First of all we note that Nos. 1, 2, 3, 5, and 6 are practically what Pascal required, as we saw on p. 91. As for dates, the *Logique* first appeared in 1662, Pascal's *Pensées* did not see the light until after his death; he died in 1662, the *Pensées* came out in 1669, and Arnauld was one of those who produced them. I do not think that we should conclude that either author copied from the other. Pascal had by far the greater knowledge of mathematics, as Arnauld freely confesses, and the *Pensées* represent what had been long on his mind. On the other hand, Arnauld would not have stressed ideas taken from a friend without some reference to their origin. The natural explanation is that the two friends worked in such close accord that it would be next to impossible to say which were the ideas of which author.

Arnauld has pointed out that the definitions of mathematics are purely nominal, and he is very scornful of those who seek to define things whose significance is perfectly clear. We have, he says, a pretty good idea what a man is, and learn nothing of value when Plato defines him as a featherless biped.† Plato acknowledges that the ideal way would be to define everything, but realizes perfectly that this would be impossible. We are entirely in sympathy with all this to-day. On the other hand, we have drawn away from his No. 3, which requires that the truth of an axiom should be perfectly evident, that is to say, from the purely logical point of view. It is the consequences, not the origins, of an axiom that interest us, in so far as we are examining the logical structure of any system.

Arnauld's requirement 6 is important. It means that in mathematics a term may have a different meaning from what applies in other connexions; we should continually check up and see in what sense a term

† Ibid., pp. 344 ff.

is being used. In No. 7 he speaks of 'the natural order', a very favourite idea of his, which I shall have occasion to attack later on.

In the following pages Arnauld develops these general ideas. First there is the question of definitions. He insists on using only such terms as are clear and easily understood. Then he takes a shot at the Greek definition of an angle. Here is Heath's translation of the Greek of Euclid I, definitions 8 and 9:

8) A plane angle is the inclination to one another of two lines in a plane which meet one another, and do not lie in a straight line.

9) And when the lines containing the angle are straight, the angle is called rectilineal.

Arnauld's translation is: 'La rencontre de deux lignes droites inclinées sur un même plan.'†

I will not take up the question of whether these two translations mean the same thing, but point out that Arnauld makes a very shrewd comment. He says that if Euclid wishes to use these words as a nominal definition of angle he has a perfect right to do so, provided that he bears this definition constantly in mind, but such is not the case. 'Il enseigne par exemple, à diviser un angle en deux. Substituez sa définition. Qui ne voit que ce n'est point la rencontre de deux lignes qu'on divise en deux, ce n'est pas la rencontre de deux lignes qui a des cotés, et qui a une base ou sous-tendante, mais tout cela convient à l'espace compris entre les lignes.' For Arnauld an angle was the space comprehended between two lines or line segments. Just where this space ends is not clearly stated, he measured it by the circular measure of the arc of any circle whose centre is the intersection of the lines and whose radii are the sides. The idea of calling an angle the space goes back to Apollonius, but is none the less unsatisfactory. He was probably unaware of the great amount of discussion that had taken place on this very point.‡

Another of Euclid's definitions to which Arnauld objects is that of a ratio. Here is Heath's translation: 'A ratio is a sort of relation in respect to size between two magnitudes of the same kind. Magnitudes are said to have a ratio to one another which are capable, when multiplied, of exceeding one another.' Arnauld's translation is: 'La raison est une habitude de deux grandeurs de même genre, comparées l'une à l'autre selon la quantité. La proportion est une similitude de raisons.' Arnauld points out that the difference of two quantities is just as much a 'habitude' as their quotient. He would distinguish between arithmetical and geometrical proportion just as we distinguish between arithmetical and geometrical series. The whole matter seems to me rather trivial. What Euclid really uses is not the idea of ratio, which is, at best, con-

† Arnauld, p. 283.   ‡ Cf. Euclid, vol. i, pp. 176 ff.

fusing, but of proportion, and this, following Eudoxus, he defines in truly masterly fashion, covering both the commensurable and the incommensurable case.

Arnauld is very scornful of those who maintain that all of our ideas of truth are derived, in the last instance, from physical experience. This frequently leads astray through false induction. He suggests that one would take it as an experimentally established fact that if an upright U-shaped tube be partly filled with water, the liquid will stand at the same height in the two arms. As a matter of fact, owing to capillary attraction, it will stand higher in the thinner tube, when the two are notably different. Or take the axiom that the whole is greater than a part. Our belief in this does not arise from the observation that a man is taller than his head, or a wood is greater than an individual tree, but from a perfectly clear idea of what is meant by the words 'whole' and 'part'. I do not know whether he believed in the existence of innate ideas, the question is in any case too metaphysical for our present discussion. He did not see the difficulty that Bolzano introduced with the study of infinite assemblages. He would probably have said that two geometrical figures which could be brought to coincidence by a rigid motion were of the same size, but an angle, under his definition, can be brought to coincide with a part of itself, by sliding the plane along one of its sides. He returns again and again to the consideration of what is contained implicitly in an idea. 'Tout ce qu'on voit claire-ment contenu dans une idée claire et distincte qu'on a du tout, enferme celle d'être plus grand que sa partie. Donc on peut affirmer avec vérité que le tout est plus grand que sa partie.'†

Here are six faults which he finds in the reasoning of the geometers:

I) *Avoir plus de la certitude que de l'évidence, et de convaincre l'esprit plustôt que de l'éclairer.* This means that didactic needs are more impor-tant than logical ones, a complete shift from Euclid's ground, but quite permissible in one whose main aim is to write a text-book.

II) *Prouver des choses qui n'ont pas besoin de preuve.* This again is an important point as anyone who has undertaken to teach elementary geometry has discovered; the difficulty is to draw the line.

III) *Démonstration par l'impossible.* He grants that such proofs are at times necessary. 'Cependant il est visible qu'elles peuvent convaincre l'esprit, sans qu'elles ne l'éclairent.' This again is good pedagogy, a beginner finds such proofs unsatisfactory.

IV) *Démonstrations tirées par des voies trop éloignées.* He objects to Euclid's proof of I. 5, the 'Pons asinorum', for bringing in a number of triangles which have nothing to do with the result. He was evidently not familiar with other known proofs.‡ He also disliked Euclid's own proof

---

† Arnauld[1], pp. 299 ff.   ‡ Euclid, vol. i, p. 252.

of the Pythagorean theorem for introducing extraneous triangles, when it can be proved otherwise, 'puisque l'égalité des carrés ne dépend point de l'égalité des triangles qu'on prend pour moyen de cette démonstration mais de la proportion des lignes'. It is interesting to compare this with Heath's view: 'There can be little doubt that the proof by proportion is what suggested to Euclid the method of I. 47, and the transformation of the method of proportion into one based on Book I only, effected by a construction and proof so ingenious, is a veritable *tour de force* which compels admiration.'[†]

V) '*N'avoir soin du vrai ordre de la nature*. C'est la plus grande faute des géomètres. Ils se sont imaginés qu'il n'y avait presqu'aucun ordre à garder.' This is the sorest point with Arnauld, he comes back to it again and again. He is convinced that there is a natural order in geometry, and that most of the difficulties which a beginner encounters arise from the departure therefrom. The book which I shall discuss presently was written largely to prove this point. If we deny his premiss, the conclusion falls to the ground.

VI) *Ne point se servir de divisions et de partitions*. This seems to be an objection that the subject is not divided into species and types. It is hard to take this very seriously; I cannot see how such subdivision would help much.

Arnauld was certainly not a mathematician of the stature of Euclid. This great teacher had many critics, both before and after Proclus, the critic *par excellence*. But the great Jansenist was an acute thinker and an honest one; his criticisms certainly give us much food for thought.

§ 3. *Les Nouveaux Élémens de Géométrie*

It is now time to consider in detail Arnauld's principal mathematical work which first saw light in 1667. He tells us in the preface an amusing story of how he came to write it. Pascal, whom he describes as 'un des plus grands esprits de ce siècle et des plus célèbres par l'ouverture admirable qu'il avoit pour les Mathématiques', had written a short essay on the beginnings of geometry. This fell into the hands of Arnauld, who was astonished at the confusion that subsisted in the work owing to the unnaturalness of the order. He said jokingly that if he had a little time, he could produce something much better. The occasion arose, and he produced the book under discussion. We find a confirmation in something written by Nicole: 'Lorsque Pascal vit l'ouvrage il condamna le sien au feu, et reconnut franchement que M. Arnauld avoit trouvé le vrai ordre naturel de traiter cette matière.'[‡]

'Ordre naturel': here lies the key to all of Arnauld's work, for as I have said, he was convinced that such an order existed, and that when

---

† Euclid, vol. i, p. 353.　　　　　　　　　　　　　　　‡ Bopp[1], p. 237.

it was once found, the demonstrations were easy. I do not personally feel that he found any such order, and if his demonstrations are easy, it is because he proceeds carefully, and avoids certain difficulties by the fatally simple expedient of introducing new axioms. Let us see what he says:

'Il ajoutoit même que cet ordre ne servoit pas seulement à faciliter l'intelligence et soulager la mémoire; mais qu'il donnoit lieu de trouver des principes plus féconds, et des démonstrations plus nettes que celles dont on se sert d'ordinaire. En effet il n'y a presque dans ces nouveaux élémens que des démonstrations toutes nouvelles qui naissent d'elles-mêmes des principes qui y sont établis et qui comprennent un assez grand nombre de nouvelles propositions.'†

I cannot see any justification for the statement that his new order had uncovered many new propositions, personally I find few. He probably did not have recourse to much bibliographical material. The whole passage is essentially what we call at the present time 'sales talk'.

What is the reason for studying geometry anyway? In Arnauld's view it is mental training. Logic does the same thing, but geometry does it better because the underlying ideas are more clearly set forth. Arnauld does not look upon the subject-matter of mathematics as in itself very important.

'C'est une ignorance tres blâmable que de ne point savoir que toutes ces spéculations stériles ne contribuent rien à nous rendre heureux; qu'elles ne soulagent point nos misères, qu'elles ne guérissent point nos maux . . . cependant on ne voit que trop par expérience que ces sortes de connoissances sont d'ordinaire jointes à l'ignorance de leur prix et de leur usage. On les recherche pour elles-mêmes, on s'y applique comme à des choses fort importantes. On en fait sa principale possession. . . . Si cet ouvrage n'a rien de ce qui mérite la réputation de grand Géomètre, au jugement de ces personnes en quoy il est très juste de les en croire, au moins on peut dire avec vérité que celui qui l'a composé est exempt du défaut de la souhaiter; que quoique il estime beaucoup le génie de plusieurs personnes qui se mêlent de cette science il n'a qu'un estime très médiocre pour la Géometrie en elle-même.'

I quote these reflections at length to give an idea of Arnauld's point of view. I cannot share Bopp's opinion that he must have had a profound knowledge of higher mathematics because he corresponded with Leibniz about philosophy.

It is time to take up the fifteen books one by one. The natural order which he praises so highly always consists in proceeding from the general to the particular. Herein he differs totally with Euclid, who begins with the simplest figure, the triangle. Arnauld's first subject of

† Arnauld², preface. The pages are not numbered.

study is 'Les grandeurs en général'. He assumes that we know what is meant by 'quantité ou grandeur générale en tant que ce mot comprend l'étendue, le nombre, le tems, les degréz de vitesse et, en général tout ce qui se peut augmenter en ajoutant ou multipliant, et diminuer en soustraiant ou divisant.' He makes certain arithmetical assumptions such as that which says that whatever we mean by signs such as a, b, c we know

$$b = b; \qquad c = c; \qquad b \times c = c \times b.$$

Then come a number of axioms, as: 'The whole is equal to the sum of its parts'; 'If equals be taken from equals, the remainders are equal'. Next come the rules for the fundamental operations when applied to letters. A polynomial is called a *grandeur complexe*. Negative quantities as such do not appear, though it is hard to believe that a correspondent of Leibniz and Descartes was ignorant of them. He is a good deal worried to show that the product of 'moins en moins donne plus'. The book ends with the simplest equations.

The second book deals with proportions. As I said on p. 106, he insists that two quantities of the same sort can be combined in two ways: we may find their difference or their ratio. And here, of course, the trouble begins, for we are never told exactly what a ratio is: 'L'autre est quand on considère la manière dont une quantité est contenue dans une autre ou en contient une autre, ce qui s'appelle raison.'†

He distinguishes conscientiously between commensurable and incommensurable quantities. Two ratios are equal when the antecedents contain or are contained in the consequents equally. This is rather a dark saying. He flounders around a good deal, gives five axioms, but finally gives a second definition: 'Deux raisons sont appelées égales quand toutes les aliquotes pareilles des antécédents sont chacunes également contenues en chaque conséquent.'

This is nothing but Eudoxus' classic definition which we find in Euclid V, definition 5. One wonders why Arnauld imagined that he was making things easier by putting this supremely difficult material into his second book, Euclid postponed it as long as he dared. If in various places Arnauld's proofs in questions of proportion are shorter than Euclid's, that is because he dares not drag all the difficulties out into the light. I think, however, it is fair to say that he did not intentionally slur over them, but rather did not understand all that was involved. The remainder of Books II and III is devoted to proportion. He becomes rather deeply involved when he passes from equal to unequal ratios; he writes:

'C'est ce qui n'est pas peu embarrassé. Mais voicy, ce me semble, la plus facile manière de la concevoir.

† Arnauld², p. 23.

'De deux raisons inégales celle qui s'approche davantage de la raison d'égalité s'appelle la plus grande et celle qui s'en éloigne la plus petite.'[†]

Here he is certainly above his depth. Presumably he is dealing only with ratios of positives, but the statement is true only of ratios of less to greater, and, as he acknowledges, it is not by any means easy to say which of two is nearest to unity. He helps out with two axioms. If we have two ratios with the same consequent, that with the greater antecedent is the greater ratio. If two ratios are equal to two others which are unequal, that is the greater which is equal to the greater ratio. He ends up pathetically: 'Mais il n'est pas toujours facile de discerner en toutes sortes de grandeurs quelle est la plus grande différence de deux termes aux termes communs.'[‡]

Book IV is devoted to commensurable and incommensurable quantities. He gives as his fundamental theorem that which states that the square or cube of a rational ratio (*raison de nombre*) is the ratio of the squares or cubes of the terms. The book ends with a few rules about integers.

Rule I) The square of every odd number which, when added to itself less unity, gives a perfect square, is the sum of two perfect squares.

Let
$$x = 2b+1, \qquad 2x-1 = 4b+1 = p^2,$$
$$x^2 = (2b)^2 + p^2.$$

Rule II)   If
$$c^2 = a^2 + b^2,$$
$$(rc)^2 = (ra)^2 + (rb)^2.$$

He then builds up, I do not know why, an elaborate table in seven columns. Each term in column V is the sum of the corresponding terms in columns II and III, each in column VII of the two in V and VI.

The fourth book of his first edition, the only one I have seen, deals with endless geometrical series, as

$$\tfrac{1}{2}+\tfrac{1}{4}+\tfrac{1}{8}+\dots; \qquad \tfrac{1}{3}+\tfrac{1}{9}+\tfrac{1}{27}+\dots; \qquad \tfrac{1}{4}+\tfrac{1}{16}+\tfrac{1}{64}+\dots.$$

He sums the latter by what amounts to the formula

$$a+ar+ar^2+\dots = \frac{a}{1-r} - \frac{ar^n}{1-r}.$$

When $|r| < 1$ the last term can be made as small as we please. Arnauld remarks that the famous sophism of Achilles and the tortoise can be explained in this way.

The subject of geometry in the narrow sense begins with Book V. This deals with extended magnitudes. The simplest of these are one-dimensional and the book deals with straight lines and circles. He assumes that the reader knows what a straight line is, and that he also

† Ibid., p. 56.   ‡ Ibid., p. 57.

knows that it gives the shortest path between two points. However, this is not enough; he must assume other axioms. He lays down six other axioms about straight lines, one being Archimedes' axiom about convex paths. The axiom that two different straight lines cannot share more than one point appears in several forms. He also makes a vague sort of parallel axiom, 'Deux lignes droites qui estant prolongées vers un même costé s'approchent peu à peu, se couperont à la fin'. Then come six axioms about the circle or circumference. One of these is really important and saves a lot of time later on. 'In the same circle or equal circles, equal chords subtend equal arcs, and conversely.' This is very helpful to him later on. Certainly there is a tendency in modern text-books to extend the list of axioms. This comes from the discovery that Euclid really makes a whole lot of assumptions which he does not explicitly acknowledge. A rigorous proof of this axiom based on an adequate definition of the length of an arc would be laborious. On the other hand, it takes a certain amount of self-control not to avoid all the major difficulties by indulging freely in additional axioms. The whole theory of incommensurables has pretty much gone out of the window in recent times, largely because the Euclidean theory is really too hard for most pupils to understand, and various substitutes are less rigorous, and not much simpler.

Arnauld next passes to oblique and perpendicular lines and gives much attention to them as affording a means of proving various theorems which Euclid handles with the aid of triangles. We have here another curious axiom. 'If two lines intersect, and if two points on the first are each equidistant from two points of the second, the same is true of all points of the first.'† This is very convenient, but quite unnecessary if we follow Euclid's order. The axiom that the straight line gives the shortest path and Archimedes' axiom about convex paths is very helpful in this book.

In Book VI Arnauld takes up parallel lines and makes, on the whole, pretty heavy weather of it. He points out acutely enough that we associate with parallels two distinct ideas, that of non-intersection and that of equidistance. Many fruitless attempts have been made to prove Euclid's parallel axiom by playing on this distinction. Arnauld's principal tool in this work is his set of theorems about perpendiculars and obliques, established in the preceding book. He also has a very strong parallel axiom.

Book VII is much more interesting; it deals with the relations of straight lines and circles. First comes the problem of passing a circle through three non-collinear points. And then, surprisingly enough, we have a little trigonometry. We are told about the sine and the versed sine of an arc, less than a quadrant. He points out that the sine can be

† Arnauld², p. 87.

taken as a measure for the supplementary arc. Then comes the important theorem that in two concentric circles the arcs which correspond
to the same central angles are proportional to the circumferences. This
is based on his axiom that equal chords subtend equal arcs, and a careful
handling of the aliquot parts. He considers as the measure of an arc
its ratio to the circumference, essentially its circular measure. He shows
that arcs of equal measure have equal sines. The book ends with a few
theorems about secants and chords and the simplest properties of
tangents.

In Book VIII we come to angles. An angle is the 'area comprehended
between two line segments with a common origin, but not on the same
line'. He saves himself much vagueness by taking as the measure of an
angle the circular measure of the arc of any circle whose embracing radii
are along the sides of the angle. Equal angles have equal sines; the
equality of the alternate interior angles made by a transversal with two
parallel lines comes from the equality of their sines; vertical angles are
equal because they intercept equal arcs.

In Book IX we study the relation of an angle to a circle whose centre
is not at the vertex. Here is the most fundamental theorem: 'Tout
angle compris entre une tangente et une corde a pour mesure la moitié
de l'arc soutenu par cette corde, du costé de la tangente.'† This is
Euclid, III. 32, but Arnauld's proof is much neater. The complementary
angle between the radius and the chord is equal to the angle between the
radius and the diameter parallel to the chord, which is complementary
to half the angle which the chord subtends at the centre. This is the
neatest proof I have seen; the usual theorems about the angles of other
chords are deduced easily enough. An excellent book.

In Book X we come to proportional line-segments. Here Arnauld
introduces a curious element, the 'parallel space'. This is the infinite
space between two parallel lines. He takes as its measure the length
of any common perpendicular. A segment which two parallel lines cut
on a transversal not perpendicular to them is called an 'oblique'. His
first fundamental theorem is that if we have two parallel spaces, the
obliques making the same angles with them are proportional to the
perpendiculars. The proof passes through the commensurable and
incommensurable cases, as one would expect. It follows at once that a
parallel to the edges within a parallel space will divide all obliques
proportionally. From this we pass easily to the fundamental theorem
that if a triangle be crossed by a line parallel to one side we have two
triangles with proportional sides. I do not rate this book very highly
except for ingenuity. Euclid's approach to the subject, which is based
on Euclid, VI. 1, 'Triangles and parallelograms which are under the

† Arnauld², p. 165.

same height are to each other as their bases', seems to me more natural. Arnauld has not yet brought in the subject of area, he substitutes therefore his parallel space which is, after all, infinite. The book ends with a proof of Euclid, VI. 3: 'The bisector of an angle of a triangle divides the opposite side proportionately to the adjacent sides'.

Book XI is still more interesting and original. Arnauld writes: 'Ce livre-cy sera encore de la proportion des lignes, et contiendra plusieurs choses nouvelles qu'on jugera, peut-estre, plus belles et plus générales que tout ce qu'on a trouvé jusque icy sur cette matière des proportions.'† The subject is *lignes réciproques*. This means, in general, that we have four line-segments such that the first is to the second as the fourth is to the third. He is working around towards antiparallelism. A transversal is anti-parallel to the base of a triangle when it makes with the other two sides b and c the angles which the base makes with c and b. This again amounts to saying that its intersections with these two sides are concyclic with the ends of the base. I find nothing corresponding to this in Euclid. He recognizes that the great interest in the whole subject comes from this last-mentioned fact. Here is his favourite theorem, which I restate, as his statement is needlessly long. Given a point P on a circumference, where Q is the other end of the diameter through P and Q', the intersection with a perpendicular on this diameter. Then if a line through P meet the circumference in R and the perpendicular in R',

$$PQ.PQ' = PR.PR'.$$

Arnauld is lyrical in his praise of this theorem: 'Voicy la proposition générale sur ce sujet, qui est, peut-estre, la plus belle et la plus générale qu'on puisse trouver sur les proportions des lignes qu'on puisse trouver par la géométrie ordinaire.'‡ The words *géométrie ordinaire* suggest his familiarity with the new analytic geometry of Descartes. His treatment is rather prolix, but he uses it to prove Euclid, III. 35 and 36, which say in effect that the product of the distances from a point to the two intersections of a circle with a line through the point is independent of the direction of the line. The book ends with a number of familiar theorems and constructions. On p. 237 he gives a pretty solution of the problem of golden section, cutting a length in extreme and mean ratio. This is simpler than Euclid II. 11 and VI. 30, but apparently not new, having been found by Hero of Alexandria, whose work Arnauld very likely never saw.§

Book XII is devoted to polygons, 'figures' as he calls them. The sum of the interior angles is $2(n-2)$ right angles. Inscriptible and regular polygons receive some notice. A circle is defined as a regular polygon of an infinite number of sides, a bad statement that lingers to this day.

---

† Arnauld², p. 208.   ‡ Ibid., p. 222.   § Tropfke, 3rd ed., vol. iv, p. 244.

Here we get from Arnauld some very shaky mathematics.† Suppose that we have two inscribed regular polygons of the same number of sides. Let b be the side of the one, c of the other:

$$b/c = 10b/10c = 100b/100c = \dots .$$

'Donc les circuits ne sçauront manquer d'estre en mesme raison que les costéz.' This is certainly very queer, for b and c are functions of the numbers of sides. He should have shown that the perimeters were proportional to the radii, a fact that he deduces herefrom. He closes by inscribing regular polygons of various members of sides, pentagon, decagon, and pentadecagon, following Euclid, IV. 16.

Book XIII deals with the angles of triangles and quadrilaterals. Here we find an old friend from Euclid, Book I. 18 and 19: if two angles of a triangle are unequal, the opposite sides are unequal, and the greater side is opposite the greater angle. A very simple proof comes by circumscribing a circle about the triangle; the greater side will subtend the greater arc and so correspond to the greater angle. This is so simple that I suspect a trap somewhere; it is another case where Arnauld makes skilful use of the relation of chord to arc. We have the usual theorems about equal triangles which are found in Euclid, Book I. Next we have similar triangles and a proof of the concurrence of the altitudes of a triangle. This theorem is not in Euclid, but was known to Archimedes and Proclus. Tropfke calls Arnauld's proof 'den besten seiner Art bis Gauss'. Arnauld follows with theorems about right triangles, parallelograms, and the gnomon. He shows that the diagonals of a regular inscribed pentagon divide one another in extreme and mean ratio; this is essentially Euclid, IV. 11.

In Book XIV we come at last to the measurement of areas. Here again are a battery of new axioms:

Axiom I] All squares of the same sides are equal.

Axiom II] Two rectangles with the same dimensions are equal.

Axiom III] The product of a whole multiplied by a whole is the sum of the products of the parts of one multiplied by the parts of the other.

It is interesting to note that the first two of these might be proved by superposition, which Arnauld wisely avoids. There follow certain algebraic identities, as

$$\left(\sum_i a_i \right)^2 = \sum_i a_i^2 + \sum_{i,j} a_i a_j.$$

Then comes the fundamental theorem that rectangles with equal bases are to each other as their altitudes. The proof runs through the usual course for the commensurable and the incommensurable cases. Important metrical theorems, as Euclid, III. 35 and 36 and the Pythagorean,

† p. 257.

as well as, curiously enough, the less important Euclid, XIII. 10: 'If an equilateral pentagon be inscribed in a circle, the square on the side of the pentagon is equal to the squares on the side of the hexagon and on that of the decagon inscribed in the same circle.' Both Euclid's and Arnauld's proofs are rather long; the matter is simpler if we write

$$\frac{d}{2r} = \sin 36°; \qquad \frac{r-x}{2x} = \frac{x}{2r} = \cos 72° = 1 - 2\sin^2 36°;$$

$$\frac{x}{2r} = 1 - \frac{d^2}{2r^2}; \qquad d^2 = 2r^2 - rx = 2x^2 + rx = r^2 + x^2.$$

Book XV also deals with areas. At the outset we find something decidedly interesting. The *Nouveaux Élémens* were published in 1657. Four years earlier appeared Cavalieri's *Geometria Indivisibilium*, which created a great stir. Arnauld writes:[†] 'Je n'ay rien veu de ce qui a esté écrit, mais voici ce qui m'en est venu dans l'esprit, en ne m'arrestant maintenant qu'à ce qui regarde les surfaces. Le fondement de cette nouvelle géométrie est de prendre pour l'aire d'une surface la somme des lignes qui la remplissent.' He sees very clearly the dangers of any such method. *Ligne* means for him, of course, segment. His first caution is that two segments must not intersect, so that when we are dealing with line segments they should be parallel, and if circular arcs, they should be on concentric circles. Secondly, he holds rightly, that in speaking of infinite sums we are playing with fire. Consequently he says, *by definition*, two infinite sums of line segments are equal in number when they are contained in equally wide parallel strips. With these definitions he shows that two parallelograms of the same height and equal bases are equivalent because they are composed of an equal number of equal segments. The following is more interesting:

Theorem] *Le cercle est égal au triangle rectangle qui a pour costéz de son angle droit le rayon du cercle, et une ligne égale à la circonférence du cercle.*

The proof is easy. The triangle is formed of a radius and a tangent whose length is the circumference. Then the circle is built of concentric circles, and the triangle from tangents to these parallel to the original tangent, and limited by the same two lines. The numbers are the same, for we have a one-to-one correspondence. It is interesting to see this reaction of an intelligent contemporary to this startling new doctrine.

§ 4. Magic squares

The last section of Arnauld[2] is devoted to what he calls *Quarréz magiques*. There have been indeed many writers on this subject to which, at first, a mystical significance was attached. I hasten to say that our own Benjamin Franklin, who contributed to this literature,

† p. 307.

did not, probably, see any religious implications. Curiously enough, those who have written on the subject pay little attention to Arnauld. In Guenther† is a statement that Arnauld must have leaned heavily on Bachet. I find no justification for this statement, nor any similarity in their methods. Bachet acknowledges that he can write no general principle for constructing squares of even order. Here are some of Arnauld's principles which seem to me original.

We shall speak of two squares, the one natural, the other magic. Each has n cells on a side, the numbers running from 1 to n^2. Two numbers, a large and a small, whose sum is $(n^2+1)/2$ shall be called complementary. In the natural square they lie symmetrically with regard to the centre. In both squares the complement of a number on a diagonal is on the same diagonal at the same distance from the centre. In every magic square of odd order, and every one of even order, except for the sixteen central cells, if a number is in the top or bottom quadrant formed by the diagonals, its complement is in the reflection of its cell in a horizontal line through the centre, the complement of a number in the right or left diagonal quadrant is in the reflection in the vertical line through the centre. The sixteen central cells of an even-order square seem to be filled by main force with regard to what is outside.

Suppose, then, we have a centrally situated square n cells on a side, and that this is magic. When n is odd the numbers in a row, column, or diagonal of this shall add up to $\frac{1}{2}n(n^2+1)$, when n is even they add to the same. Suppose then we have such a centrally situated square. We fill the top row and right column bordering with any numbers we please not yet used, and fill the bottom and left column with their complements as already explained; the new central square fulfils the requirement. Continuing in this way the whole square is filled out.

There are a number of details in the description which I have omitted for the simple reason that I do not understand them. Arnauld is at times obscure, and does not always carry out in practice what he says. Bopp says: 'Wir werden hier nicht die eleganten Beweise Arnaulds, welche auf Analysis situs gegründet sind, wiederholen; sondern verweisen auf den französischen Text.'‡ I confess I do not find any traces of analysis situs; Bopp's arithmetical treatment covers twelve pages.

§ 5. Final estimate

What shall be our final judgement of Arnauld, and what he has tried to accomplish? Here is his own judgement of himself:

'Je laisse d'autres problèmes qui sont très faciles à résoudre par les principes qui ont esté établis. Outre que n'ayant entrepris ces Élémens que pour donner un essay de la vraie méthode qui doit traiter les choses simples

† q.v., p. 232.   ‡ Bopp[1], p. 323.

avant les composées, et les générales avant les particulières, je pense avoir satisfait à ce dessein, et avoir montrez que les géomètres ont eu tort d'avoir négligé cet ordre de la nature en s'imaginant qu'ils n'avaient autre chose à observer, sinon que les propositions précédentes servissent à la preuve des suivantes ; au lieu qu'il est clair, ce me semble, par cet essay, que les élémens de la géométrie, estant réduit à leur ordre naturel, peuvent estre aussi solidement démontrez, et sont, sans comparaison, plus aisez à concevoir et à retenir.'†

Arnauld bases his whole claim on the virtues of his new order. How far have his successors acknowledged their debt to him ? Bopp[1] points out‡ that Lamy and Varignon copied him closely, the former speaking of Arnauld as the man who had found the only real order for geometrical theorems. That curious volume *Élémens de géométrie de Monseigneur le Duc de Bourgogne*, which is based on the teaching which M. de Male-zieu delivered to that exalted personage, follows Arnauld's lead. It is also noteworthy that the book was noticed by the *Philosophical Transactions* and the *Journal des Sçavans*. There were several editions after the first. Yet to a modern teacher the work is not satisfactory either mathematically or pedagogically. It is also highly significant that when Arnauld's distinguished compatriot Adrien Marie Legendre prepared an introduction to geometry which was to have a very great influence all over Europe and America, he took Euclid and not Arnauld as his model.

Why, then, have I paid so much attention to Arnauld ? It is for his priority in breaking away from Euclid, a most important step. He tells us in his preface : 'Car tant de personnes ont demandéz au Libraire une nouvelle géométrie qu'on n'a pas pu la refuser aux instances qu'il a fait de leur part pour l'obtenir, n'estant pas juste de se faire beaucoup prier, pour si peu de chose.' We have in Kokomoor[1] and Kokomoor[2] the list of a large number of attempts that were made to improve the teaching of elementary geometry in the seventeenth century, but such examination as I have made of them has yielded very little result. Pierre de la Ramée (q.v.) throws in new axioms whenever he wants to, is careless in many proofs, and avoids all difficulties arising from incommensurables. His great interest is in practical applications, and he wishes to open an easy road to them. His work was written before Arnauld's, and I can see no reason to believe that Arnauld was influenced by him. The same is true of Robert Recorde's *Pathway to Knowledge*, which is even more casual. Professor Karpinski of the University of Michigan, a man very well versed in such matters, wrote me in April 1945 : 'I think it safe to conclude that Arnauld made the first serious attempt.' It was a very important forward step in education ; I am glad that my last word about Arnauld should be so much to his praise.

† Arnauld², p. 323.   ‡ pp. 245 ff.

CHAPTER IX

JAN DE WITT

§ 1. Geometrical writings

MOST of the men whom I have classified as amateurs in mathematics gave the remainder of their professional lives to quiet intellectual pursuits. Perhaps it would not be quite accurate to say this of Leonardo da Vinci, who gave some attention to waging war, but even in his case the balance of intellectual activity went to affairs of the mind, quiet pursuits of the studio or of the study. This, however, is far from being true of that great patriot, the Grand Pensionary of Holland and West Friesland, to whom I now turn. A very active man was he, in war and peace, deeply involved in international affairs and the internal politics of the United Provinces. His life was busy, contentious, and successful, but his enemies were determined, *and his end was tragic. It is truly surprising that such a man should find any time to devote to pure mathematics.

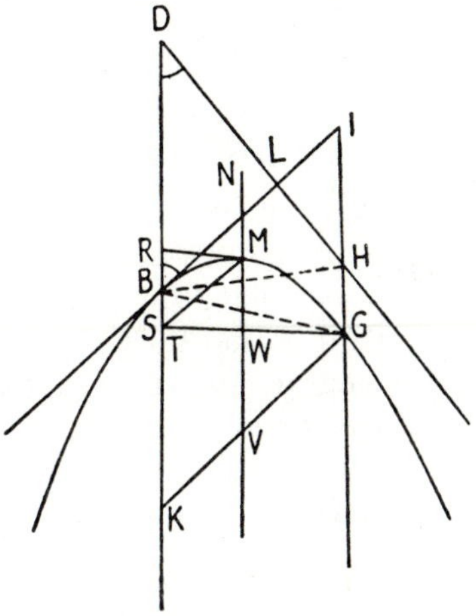

FIG. 47

The explanation of all this is to be found in the fact that De Witt, when a student of jurisprudence, lived in the house of Franz van Schooten, who was indeed professor of that subject, but also a keen mathematician and a correspondent of Descartes. In fact he published so much mathematics that he should be classed as a professional mathematician, which is not the case with De Witt, whose writings on geometry present evidence of the influence of his older friend.

De Witt explains his point of view in the preface to De Witt[1] which was written in 1658 and published the next year. He acknowledges that the methods developed by the ancients were adequate for handling plane loci, that is to say, straight lines and circles, and that the recently discovered analytic methods are applicable to loci of every sort. He was convinced that the study of curves, especially the conic sections, as sections of a cone or other surface represented a needless detour. The natural approach is a kinematical one, to study a curve as the locus of a point moving according to a mathematically described law. The first book of De Witt[1] is devoted to exactly such a study

---

* 'his end was tragic'. He was shot, publicly hanged, and his body violated by an angry mob during the panic that swept Holland when it was simultaneously invaded by France and England.

of the conics, and is sent with affectionate greeting to *amicissimus Schootenus*.

De Witt begins with a parabola which he defines in a manner which I have not seen in the writings of any previous mathematician. Suppose (Fig. 47) that we have an angle of fixed magnitude $\angle HBG$ which rotates about a fixed vertex B. Let the angles $\angle LBD$ and $\angle LDB$ both be equal to $\angle HBG$. Let H slide along a fixed line DL, and through it draw a line HI parallel to BD. Then G, whose locus we seek, is the intersection of this with the swinging arm BG. We also let GK be parallel to LB where K is on BD.

$$GK = BI = DH; \qquad \angle DBH = \angle IBG;$$

$$\frac{BD}{DH} = \frac{BI}{IG} = \frac{BD}{KG} = \frac{KG}{KB}; \qquad KG^2 = BD.KB.$$

De Witt concludes: 'Constat itaque curvam intersectione, uti prae-dictum est descriptam, eam ipsam esse, quae Veteribus Parabola.'

It is clear that BL will be the tangent at B, but the trouble with such a generation is that the point B would seem to play a special role in form-ing the curve, and that is intolerable. We must show how to replace it by any point M on the curve. I abridge De Witt's reasoning somewhat.

Let us take MS parallel to the tangent at B and R on BD so that $SB = BR$. We draw the diameter through M and let it meet GK in V:

$$MS^2 = VK^2 = DB.BS.$$

Draw GWT parallel to MR, and on the diameter through M make $MN = MR^2/BS$.

$$GK/TK = MS/RS = DB/2MS = DB/2VK,$$

$$2GK.VK = TK.DB;$$

$$GK^2 = DB.BK;$$

$$GV^2 = (GK-VK)^2$$
$$= DB(BK-TK+BS)$$
$$= DB.RT = DB.MW;$$

$$GW^2 = GV^2.MR^2/MS^2$$
$$= MW.MN.$$

This gives the desired equation of the parabola with the new parameter.

I said that I have not seen this method of generation in the work of previous geometers, I add that I am a little surprised at De Witt's use of it. In van Schooten[2] we have the description of a very simple mechanism for drawing the curve. This is based on the fact that any point on the extension of the diagonal of a rhombus is equidistant from the ends of the other diagonal. One vertex of a jointed metal rhombus is a fixed pivot, the opposite vertex slides along a fixed grooved rod, while a perpendicular to this rod at that point is brought to intersect the other diagonal. We have here a moving point whose distance from a fixed point is equal to its distance from a fixed line, so that it will trace a parabola. This, I say, is given in van Schooten[2],† which was published the year before De Witt sent in his paper, and it appears in

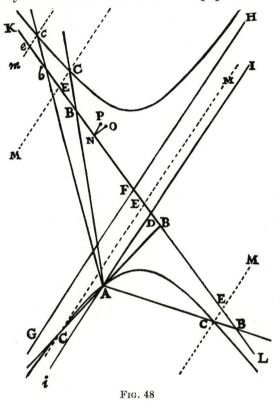

Fig. 48

the preface that van Schooten was familiar with this instrument some years before. It is hard to believe that De Witt did not know of it, or that he shared the Greek prejudice against anything which could be accomplished mechanically. I note also the requirement that

$$\angle LDB = \angle LBD$$

seems a bit extraneous. When it is removed the locus‡ is a hyperbola. De Witt judged rightly that this method of determining the curve is inferior to the one I will now explain.

Let a line segment of fixed length EB slide along a fixed line FL. Through E we pass a line EM having a fixed direction. A is a fixed point; we bring AB to meet EM in C, whose locus we seek.

† p. 357.   ‡ De Witt[1], pp. 231 ff.

When AC takes the given direction, B shall fall at D and E at F. Through F draw FH in the given direction:

$$\frac{CE}{EB} = \frac{AD}{DB} = \frac{AD}{EF},$$

$$CE.EF = EB.AD.$$

Clearly C can come as close to FL or to FH as we please without

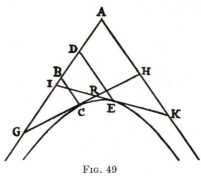

Fig. 49

reaching either. The product of the distances from C to each line in the direction of the other is constant. This is the fundamental relation of the points of a hyperbola to the asymptotes.

What about the discovery of this method of reaching the curve? I am afraid that we cannot give the credit to De Witt, for it is set forth in the writings of Descartes, and the corresponding equation determined, and De Witt was familiar with these.† A machine for describing hyperbolas based on this is described on p. 332 of van Schooten[2].

As the product of the distances to the asymptotes in two specified directions is constant, so is the product in any two given directions. In particular, if we take two general points of the curve, the product of the distances to the asymptotes along the line joining them is the same. Consequently the two longer segments on this line from the curve to the asymptotes are equal, as are the two shorter segments. A limiting case of this is that the two distances along a tangent from the point of contact to the asymptotes are equal.

De Witt proves the constant area of a triangle bounded by a tangent and the asymptotes very easily as follows:

$$\triangle GBC = \tfrac{1}{4}\triangle GAH; \qquad \triangle DIE = \tfrac{1}{4}\triangle AIK.$$

If we draw through C and E lines parallel to AG and terminated in AK, their lengths would be AB and AD; hence

$$GB.BC = AB.BC = AD.DE = ID.DE$$

$$\triangle GAH = \triangle IAK.$$

† Descartes[1], p. 18; Descartes[2], p. 22.

We find what we might call the Greek form for the equation of the

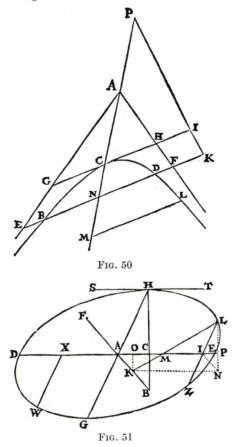

Fig. 50

Fig. 51

hyperbola very easily from this (Fig. 50). C and P are the ends of a diameter:

$$CH^2 = DF.DE = (NF-ND)(NF+ND),$$

$$ND^2 = NF^2 - CH^2,$$

$$\frac{NF}{CH} = \frac{AN}{AC},$$

$$ND^2 = \frac{CH^2}{AC^2}(AN^2 - AC^2)$$

$$= \frac{CH^2}{AC^2}.PN.CN.$$

This whole approach to the hyperbola seems to me excellent, the best thing in De Witt's geometrical work.

In the third chapter of Book I De Witt turns naturally to the ellipse, and applies a still different technique. Suppose that we have a rod of fixed length KM which slides with its ends on two fixed lines AF, AP. We choose a fixed point L on the rod, or rather on its extension, and seek its locus. One position of the rod will be BC when it is perpendicular to AE; L will then take the position H. He calls $GH = 2AH$ the 'secant'.

Through K and L draw KO and LP parallel to BC and terminating on AE. Draw LI parallel to HA, and IN parallel to AB where N is on LP

$$\frac{BA}{KA} = \frac{BC}{KO} = \frac{KM}{KO} = \frac{ML}{LP} = \frac{HC}{LP} = \frac{BA}{IN} = \frac{AC}{IP};$$

$$KA = IN.$$

Hence $AKIN$ is a parallelogram, $KN = AI$

$$AD = AE = KL$$

$$DI.IE = AD^2 - AI^2$$

$$= KL^2 - KN^2$$

$$= LN^2$$

$$\frac{LI^2}{DI.IE} = \frac{LI^2}{LN^2} = \frac{AH^2}{HB^2} = \frac{AH^2}{AE^2}$$

$$LI^2 = \frac{AH^2}{AE^2}.DI.IE$$

$$y^2 = \frac{b^2}{a^2} x(2a - x).$$

Here we have the classical Greek form for the equation of the ellipse where the axes are a diameter and the tangent at its end.

Now for the question of priority. We find in Proclus† this:

'Nor yet if you suppose a right line moving in a right angle and by bisection to describe a circle, is the circle on this account produced with mixture ? For the extremes of that which is moved after this manner, since they are equally moved, will describe a straight line, and the centre, since it is equally developed, will describe a circle; but other points an ellipse.'

Here we have the theorem quaintly enough stated when the lines are mutually perpendicular. I do not know whether De Witt was original in proving it when they make any angle, but should be very sceptical on the point. When the lines are perpendicular, a point rigidly connected with the rod, whether on its line or not, will describe an ellipse.

† q.v., p. 130.

In van Schooten[2][†] we have the description of quite a number of simple ellipsographs depending on this principle. The credit here seems to go back to Nasir Edin.[‡]

Returning to De Witt we see, by symmetry, that AE bisects the chords parallel to AH which will be parallel to the tangents at D and E:

$$DI \cdot IE < AE^2; \qquad LI^2 < AH^2.$$

It follows from this that a line through H parallel to AE cannot enter the ellipse, and so will be the tangent. Thus the diameter AE is conjugate to AH or the relation of conjugate diameters is a mutual one. An ellipse will be completely determined by any pair of conjugate diameters. In De Witt[1][§] we have a long and elaborate discussion of this problem: 'Given a pair of conjugate diameters of an ellipse, to find the conjugate of any diameter.' Here we find a good deal of waste motion. He proves a few pages later that the ends of a diameter are harmonically separated by the intersection with a chord in the conjugate direction and the intersection with the tangents at its extremities, a theorem known to Apollonius. This enables us to draw a tangent at any point and so find the conjugate to any diameter. In the fourth chapter of the first book De Witt describes certain other less simple constructions, which call for no further comment.

Book II of De Witt's *Elementa curvorum* is analytic, and devoted to showing that any linear equation represents a straight line, and any quadratic one a circle or a conic section; that is to say, if there is any corresponding locus at all. This is by no means new. Fermat did essentially the same thing in 1643; Descartes solved the analogous problem of four lines in 1637.[||]

De Witt first shows that an equation which we should write

$$y = \pm lx \pm b$$

always represents a straight line. I should mention here that no negative quantities are ever involved, a point I shall return to in the next chapter. Next we have an elaborate discussion of the parabola. He shows that by a change to parallel axes we can reduce the equation

$$y^2 = \pm ax \pm b^2$$

to the simpler form $y^2 = 2mx.$

Here is something more complicated:[††]

$$y^2 - \frac{bxy}{a} = -\frac{b^2x^2}{4a^2} + bx + d^2.$$

I mention in passing that De Witt retains the classical requirement

† pp. 303 ff.   ‡ Coolidge[1], p. 152.   § pp. 214 ff.
|| Fermat, q.v., vol. i, pp. 90 ff.   †† De Witt, vol. ii, p. 263.

that equations shall be homogeneous. That is because they are always conceived in geometrical terms. If linear, lengths are involved, if quadratic, areas. Descartes exploded this superstition by showing that all symbols represented numbers. Thus the symbol x^2 does not represent a square figure, but the fourth term in the numerical proportion

$$1:x::x:x^2.$$

De Witt first writes his equation:

$$\left(y-\frac{bx}{2a}\right)^2 = bx+d^2.$$

Put
$$y = z+\frac{b}{2a}x; \qquad z^2 = bx+d^2.$$

The question is, What geometrical significance shall we attach to these symbols? z is the distance in the y-direction to the intersection with

$$y = \frac{b}{2a}x.$$

This shall be the new x'-axis, the new origin shall be the intersection of the curve with this line. The distance along the x-axis to the old origin shall be k. We have

$$\frac{k}{x'} = \frac{d^2}{bx+d^2}; \qquad z^2 = \frac{d^2}{k}x'.$$

Here is a typical parabola. In chapter iii we handle the central conics in the same way. Our standard form is

$$y^2-c^2 = \frac{b}{q}x^2, \qquad x^2 = \frac{q}{b}(y^2-c^2).$$

Put
$$y+c = y'; \qquad x^2 = \frac{q}{b}y'(y'-2c).$$

A typical hyperbola. For the general case we take

$$y^2+2\frac{b}{a}xy+2cy = \frac{b}{a}x^2+bx+d^2,$$

$$z = y+\frac{b}{a}x+c,$$

$$z^2 = \frac{b^2+ab}{a^2}x^2+\frac{ab+2bc}{a}x+c^2+d^2,$$

$$\frac{a^2}{b^2+ab}z^2 = x^2+\frac{a(ab+2bc)}{b^2+ab}x+\frac{a^2(c^2+d^2)}{b^2+ab}.$$

Geometrically speaking we have preserved the direction of the x-axis, but changed the y-direction to $y = -bx/a$. Putting

$$x + \frac{a}{2}\frac{ab+2bc}{b^2+ab} = u,$$

a typical hyperbola, $$\frac{a^2}{b^2+ab}z^2 = u^2 + k.$$

In chapter iv De Witt takes up the same problems in greater detail, but I cannot see that he adds anything essentially new. In general I cannot count his geometrical work as of first importance. The idea of studying the conics dynamically as generated by moving points connected with moving lines is fundamental, but was much better done later by MacLaurin, and put in final form by the Chasles–Steiner construction through projective pencils. I do not agree with some writers who look upon him as anticipating that. He seems to have no knowledge of the work of Desargues and Pascal, and his discussion of the general equation marks no great advance.

§ 2. Annuities

De Witt, a forward-looking statesman, was interested in various questions of public importance. One of these was the subject of annuities. Such things had been in existence since Roman times, but the theory had been little developed. It is generally stated that De Witt was the first to make a careful study of the underlying problem, for which reason it is well to look into what he wrote on the subject. His interest was in annuities as related to public finance, and only secondarily as a means for persons with no natural heirs to secure a competence against old age. Thus all of his calculations are for annuities beginning at the commencement of a healthy normal life, let us say four years of age. His writings are in a series of letters addressed to the States-General which are available in De Witt[2] and in his correspondence with Hudde which we find in De Witt[3]. I note in passing that 1671, the date of the former, was seventeen years after Pascal's correspondence with Fermat on the subject of mathematical probability. It is not clear to me that De Witt was familiar with the studies of this discipline which were being made at the time by certain French mathematicians, especially in connexion with certain games of chance. De Witt's object was essentially practical; he does not seem to me to have penetrated very deeply into the underlying philosophy of probability. I should, however, in all fairness give the contrary opinion of his commentator:

'De Witt first gives the rules of the calculus of chances by simple examples with their explanation, that is to say, to as great an extent as is necessary for his object. They are the rules which Huygens had already developed in his treatise, but the application of the rules to the calculation of the value of life annuities is the work of De Witt, and he is, in my opinion, the first who makes so useful an application of the doctrine of chances. The law of probability was then almost unknown, at that time one could only base such calculations upon more or less probable suppositions. If we consider the slender progress which the calculus of probability had then made, we must look upon the whole treatise as an eminent proof of the inventive genius of the celebrated author.'[†]

De Witt begins by laying down certain fundamental assumptions:

'I presuppose that the real value of certain expectations or chances of objects of different values must be estimated by that which we can obtain from equal expectations or chances dependent on one or several equal contracts.'[‡]

This seems to me rather a blind statement, but it is really the germ of the theory of mathematical expectation. We learn here that if a man have the probabilities $p_1, p_2,..., p_n$ of acquiring the sums $s_1, s_2,..., s_n$ his expectation is, by definition, $\sum_i p_i s_i$.

This is what he should pay for the privilege of entering the transaction. The simplest proof is to assume that the probabilities are rational, $p_i = r_i/N$ where the denominator is very large. His expectation is $1/N$th of what he might expect to gain in N trials. Here he would expect to receive the sum s_i in what we might call asymptotically r_i times, so that the total sum gained would be $\sum_i r_i s_i$.

The expectation is, then,

$$\frac{1}{N} \sum r_i s_i = \sum \frac{r_i}{N} s_i = \sum p_i s_i.$$

De Witt does not give anything as definite as this; he begins with a very simple case. A man has a half-chance to receive 20 stuyvers and a half-chance to receive nothing. This, he says, is just the same as if a friend had put 10 stuyvers into the pool and they then tossed up to see which should take the whole.

Second proposition] 'In any particular year of life a man in full vigour is equally likely to die in the first or second half.' This is reasonable, though accurate mortality tables will not quite back it up. He extends this to assume the chance of mortality in all the half-years of a man's vigorous life is the same, an even more doubtful assumption.

Third proposition] In the age 53 to 63 the chance that a man shall survive a half-year is $\frac{2}{3}$ of what it was earlier, in the years 63 to 73 it is $\frac{1}{2}$, and in the years 73 to 80 it is $\frac{1}{3}$. We find on p. 85 of De Witt[3] this interesting statement:

'These three articles being presupposed, we have by demonstrative calcula-

† De Witt[3], pp. 95, 96.   ‡ De Witt[2], p. 82.

tion mathematically discovered and proved that the redeemable annuity being fixed at 25 years purchase as above, the life annuity should be sold at 16 years purchase and even higher to be on an equality with the other so that in the purchase of one florin of life annuity on a young and vigorous life, more than sixteen should be paid.'

This means that a life annuity of 1 florin a year for a vigorous four-year-old is worth more than 16 florins which is the paid-up price for a 25-year annuity certain, the rate of discount being 4 per cent. This I have verified. For this reason it is somewhat strange to read three pages later:

'It is likewise more useful for private families, who understand economy well, and know how to make good employment of their surplus in augmenting their capital, to improve their money by life annuities than to invest it in redeemable annuities, or at a rate of 4% per annum, because the above mentioned which are sold, even at the present time, at 14 years purchase, pay, in fact much more than redeemable annuities at 25 years purchase. I have consequently respectfully submitted to your Lordships the unchallengeable proof of my assertion.'

The surprising thing is that life annuities should have been purchasable at 14 years' purchase of a 25-year annuity certain when the latter was worth 16 years' purchase. Perhaps, in fact, attention was paid to the fact that the annuitants were above 4 years of age.

Let us see how De Witt worked this out. There is an explanation in a footnote to p. 102 of De Witt[2], and a different one in the article on annuities in the *Encyclopaedia Britannica*. The following seems to me simpler and nearer to De Witt's thought. He limits himself to 154 half-years beginning with the age 3 or 4, and kills off all who are older than 75 or 76. Suppose, first, that the probability of death is the same for every half-year in the interval. This amounts to saying that the probability that the last payment shall be in any particular half-year is the same for all 154. If then we calculate the initial worth of an annuity certain whose last payment shall be in any one of these half-years, and divide by 154 we get the average amount paid, and so the proper value for the life annuity. But De Witt does not assume equal probability for all last payments. It is the same for the first hundred half-years; for the next 20 years he assumes that the chance is $\frac{2}{3}$ of what it was before, in the next 20 it is $\frac{1}{2}$, and in the last 14, $\frac{1}{3}$. He therefore gives the first 100 sums each the weight unity, the next 20 the weight $\frac{2}{3}$, the next 20 the weight $\frac{1}{2}$, and the last 14 the weight $\frac{1}{3}$, and divides by 128, getting the value 16·001606.

De Witt gives no hint as to the mathematical procedure employed in calculating these sums; probably he did not make the calculations himself. However, the calculation of an annuity certain at a specified

rate of discount to run for a specified time involves nothing more
recondite than summing a geometric series, and that is found in Euclid,
IX. 35, and was presumably well known. Besides, such calculations
would be unsuitable matter to put into a communication addressed to
the 'Noble and Mighty Lords of the States General'. In the concluding
pages of De Witt[2] we find a number of reasons why a life annuity is
really worth more than 16; these are of no mathematical interest.

The Grand Pensionary returns to the subject of life annuities in
De Witt[3], but this time he does not assume anything so simple as an
equal probability of decease in each half-year. He assumes that a
mortality table has already been set up in this shape. We are told how
many years pass from the beginning until the first death occurs, say at
the end of r_1 years. Let the second death occur r_2 after the beginning,
and so on, the last occurring after r_n years. Let A_n be the value of an
annuity certain to run for r years. He assumes that a particular person
is just as likely to be one of these as another, hence the value of a life
annuity should be

$$\frac{A_{r_1}+A_{r_2}+\ldots+A_{r_n}}{n}.$$

I do not find the reasoning perfectly convincing, but the answer is
correct if calculated by more modern assumptions.

De Witt next proceeds to a more novel and difficult calculation, that
of an annuity based on the last survivor of two, three, or more lives.
His reasoning is similar to that which precedes. Let us start with the
same life table, the first death coming after r_1 years, the second after r_2
years, and so on. He assumes it is equally likely that the first two to
die should be any two of these. If the decedents were the first two, the
annuity would be worth A_{r_2}, if the second death were after r_3 years, the
first might equally well be after r_1 or r_2 years, so we have two equal
probabilities. If the second were to die after r_4 years, the first might
be after r_1, r_2, or r_3 years. He thus calculates the annuity of the second
to die as

$$\frac{A_{r_1}+2A_{r_2}+\ldots+(n-1)A_{r_n}}{1+2+3+\ldots}.$$

Again, suppose he with an annuity of the third to die, is the last survivor
of three. The principle is the same. If the last survivor dies at the end
of r_3 years the value is A_{r_3}. If he dies at the end of r_4 his two predecessors
were equally likely to die after r_1 years and r_2 years, or after r_1 years
and r_3 years, or after r_2 years and r_3 years, three possibilities. If he die
after r_5 years we have six equal probabilities; the answer is

$$\frac{A_{r_3}+3A_{r_4}+\ldots}{1+3+\ldots}.$$

The form in which De Witt gives his answer is interesting:

7	15	24	33	41	50	59	68
	1	2	3	4	5	6	7
		1	3	6	10	15	21
			1	4	10	20	35
				1	5	15	35
					1	6	21
						1	7
							1

The first row gives, presumably, the years of death, reduced according to some system; the remainder of the table is the miscalled Pascal triangle.

We are very safe in saying that this represents De Witt's most important contribution to mathematics, and that it is a very commendable performance for a man deeply involved in the affairs of state. But I cannot help feeling astonished that such an important subject should have to wait until his time for adequate and careful discussion.

CHAPTER X

JOHANN HEINRICH HUDDE

§ 1. The reduction of equations

It is greatly to the credit of van Schooten that two of his pupils who attained high position in public life should have spent part of their leisure in mathematical study and produced work worthy of comment. I do not know the name of any other teacher who produced two such pupils, and both remained loyal to their master, which has not always been the case. One of these pupils was De Witt whom we have just considered, the other was the man whose name stands above. He began as a student of jurisprudence and certainly retained an interest in that branch all of his life. He was appointed to take charge of the flooding of the country when it was threatened with invasion by the French, but his greatest public distinction was that he was elected Burgermeister of Amsterdam no less than nineteen times. His interest in mathematics was very real. He was personally acquainted with Huygens, who was surprised at the amount of Hudde's unpublished mathematical notes. We see from the letters in De Witt[3] that the Grand Pensioner rated him highly, and Collot d'Escury quotes one Witsen as calling Hudde the 'incomparable mathematician'.† We need not go quite as far as that, and we should not be too much prejudiced by the statement that he boasted he could pass a curve through any number of points.

Our writer's first article, Hudde[1], has to do with the reduction of equations. This is rather unsatisfactory from a modern point of view, for there is no discussion of domains of rationality, as we should expect to-day. Hudde had never heard of anything of the sort. He writes his equation by setting equal to 0 a homogeneous polynomial involving such a variable as x.

He begins with several rules which amount to this:

Suppose that we have a polynomial, homogeneous in the variables $x, a, b, c, \ldots$. Let us give to some of these variables special values. If in the resulting polynomial the term independent of x is not 0, and if it is not factorable, then the original one was not. If the term independent of x does vanish, but if the polynomial is unfactorable after x has been divided out, then the original polynomial was unfactorable. Here are two of his examples:

(i) $x^3 - (3a+b)x^2 + (2b^2+3ab+4a^2)x - (a^3+5a^2b+4ab^2+b^3) = 0.$

Let $a = 0$: $x^3 - bx^2 + 2b^2x - b^3 = 0.$

† De Witt[3], p. 96.

This is not rationally factorable, hence the original was not.

(ii) $x^4+4cx^3+(4c^2-d^2-2b^2)x^2-4b^2cx+(b^4-b^2d^2) = 0.$

Let $b = c = d = 1$:

$$x^4+4x^3+x^2-4x = 0,$$
$$x^3+4x^2+x-4 = 0.$$

This is unfactorable, hence Hudde concludes the original polynomial had no factor of degree less than three. It seems to me that he has gone wrong at this point. His next phrase is: 'At vero si ultimus terminus evanescat atque etiam inde resultans aequatio non existat reducibilis, aequatio proposita ad pauciores dimensiones quam ista resultans reduci non poterit.'†

But if we take $x^3-bx^2+a^2x-ab$

and put $b = 0$, $x^2+a^2 = 0;$

yet the original equation had the factor $x-b$.

Hudde frequently introduces new variables. For instance,‡

$$x^5+(a^2+b^2)x^3+(3ab^2-2a^3)x^2+2ab^4 = 0.$$

Let $x^3+yx^2+2a^2x+z = 0,$
$$x^3 = -yx^2-2a^2x-z,$$
$$[3a(ay+b^2)-y(y^2+b^2)-2a^3-z]x^2+[yz+2a^4-2a^2(y^2+b^2)]x+$$
$$+[(a^2-b^2-y^2)z+2ab^4] = 0.$$

Setting the coefficients separately equal to 0, and eliminating z twice,

$$-y^4+(a^2-b^2)y^2+(3ab^2-2a^3)y+2a^2(a^2-b^2) = 0,$$
$$y^5+(2b^2-4a^2)y^3+(2a^3-3ab^2)y^2+(3a^4-4a^2b^2+b^4)y+$$
$$+(5a^3b^2-ab^4-2a^5) = 0.$$

These will have a common root $y = a$, giving

$$z = 2ab^2;   x^3+ax^2+2a^2x+2ab^2 = 0.$$

I can find only one other case, and that was even more trivial, where Hudde introduced two new variables. I consequently cannot help wondering at Cajori's statement, 'With Hudde we find the first use of three variables in analytic geometry'.§

§ 2. Equal roots

The remainder of Hudde's algebraic work seems to me very prolix, not to say boring, with one important exception. On p. 435 of Hudde[1] he finds a necessary and sufficient condition that an equation shall have two equal roots. No proof is given in that place; I will give in somewhat abbreviated form the proof that he gives on pp. 507, 508 of Hudde[2].

† Hudde[1], p. 408.   ‡ Ibid., p. 428.   § Cajori[1], p. 180.

Let an equation have two roots equal to y. It may then be written

$$(x-y)^2[x^{n-2}+a_1 x^{n-3}+a_2 x^{n-4}+...] = 0.$$

Here are three successive terms:

$$(a_k-2a_{k+1}y+a_{k+2}y^2)x^{n-k+2}+(a_{k+1}-2a_{k+2}y+a_{k+3}y^2)x^{n-k+3}+$$
$$+(a_{k+2}-2a_{k+3}y+a_{k+4}y^2)x^{n-k+4}.$$

Let us multiply these in turn by three successive terms of any arithmetical progression, say $\alpha-\beta$, α, $\alpha+\beta$, and seek the total coefficient of a_{k+2} which we may take as a general term in a, and which appears only in these three. We find

$$3\alpha[x^{n-k+2}y^2-2x^{n-k+3}y+x^{n-k+4}].$$

This will vanish when $x = y$. He finds thus as a necessary and sufficient condition that an equation should have two equal roots, that if we multiply the successive coefficients, including those which are equal to 0, by successive terms of any arithmetical series, the resulting equation shall share a root with the original one. This is easy to determine by Euclid's method of finding the greatest common divisor. As a matter of fact he usually takes the series n, $n-1$, $n-2$,..., and this amounts, little as he knows it, to requiring that a polynomial shall share a root with its derivative. We shall meet this again when we come to consider L'Hospital.

We find a further discussion of equal roots when we come to the discussion of maxima and minima in Hudde[2], for a maximum or minimum usually comes at the top or bottom of an arch, or where the curve meets a horizontal line two of whose intersections fall together. More briefly we find a maximum or a minimum for y by finding such a value for x that two roots in y coalesce. He gives in fact some maxima and minima problems which involve several variables. Here is one:[†]

$$y^3-nyx+x^3 = 0; \qquad v-x = y; \qquad \tfrac{1}{2}v-y = z;$$

v to be a maximum. Eliminate y:

$$v^3-3v^2x+3vx^2 = nx(v-x).$$

Eliminate x:

$$\tfrac{1}{4}v^3+3vz^2 = n\left(\frac{v^2}{4}-z^2\right),$$

$$\tfrac{1}{4}v^3-\frac{n}{4}v^2+3z^2v+nz^2 = 0.$$

Take the series 3, 2, 1, 0:

$$3z^2 = \frac{nv}{2}-3\frac{v^2}{4}.$$

Substituting for z^2:

$$v^2 = \frac{n^2}{3}.$$

† Hudde[2], p. 514.

Hudde has here an ingenious method of finding out when an equation has two equal roots. It is his best known contribution to mathematics, and deserves more than passing notice. It was unfortunate for him that he did not discover it sooner, as it would not have been so quickly superseded by the superior methods of the infinitesimal calculus. There remains, however, one very interesting question about his place in mathematical history: Are we not indebted to him for a very important contribution to algebra, the use of a single letter to stand for a negative as well as for a positive quantity? This view seems to be widely held. For instance:

'With Descartes a letter represented always a positive number. It was Johann Hudde who in 1659 first let a letter stand for a negative as well as a positive value.'[†]

Or again:

'Den Schritt, einem und demselben Buchstaben positiven und negativen Wert zu verleihen, tat 1657 Johann Hudde, Amsterdam, bei Descartes hat noch jeder Buchstabe nur positiven Wert.'[‡]

I think that this idea was first put forth by Ennestrom (q.v.) in one of his customary attacks on the accuracy of Cantor. He points to Hudde's Regula XI which says:

'Brevitatis causa quantitatem cognitam 2^{di} termini adfectam suis signis $+$ & $-$ vocabo p, $3^{tii}q$, $4^{ti}r$, $5^{ti}s$ atque sic deinceps & $-p$, $-q$, $-r$ etc. easdem quantitates designabunt, sed contrariis signis adfectas.'[§]

This certainly means that a single letter can stand for a positive or negative quantity, a tremendous advance. But it comes after 30 pages where we continually find literal coefficients with $+$ or $-$ signs prefixed, and the same in Hudde[2] which is more recent. This would seem completely meaningless unless the letter alone meant a positive quantity. I conclude that Hudde made a most valuable mathematical discovery, which he did not in the least appreciate himself.

† Cajori[1], p. 178.   ‡ Tropfke, 3rd ed., vol. ii, p. 100.   § Hudde[1], p. 431.

CHAPTER XI

WILLIAM, VISCOUNT BROUNCKER

§ 1. Continued fractions

WILLIAM, second Viscount Brouncker, held a more exalted position in the world's eye than any other great amateur mathematician. He inherited his Irish peerage from his father, for whom it had been created, and who stood high in Court favour. William fully maintained the family loyalty to the throne. He was long intimate with Pepys, and held exalted, if not perhaps arduous, positions both in the Treasury and the Admiralty. But his great public service consisted in being one of a small number of scientific enthusiasts who founded the Royal Society. He not only helped to found it, but was the first President, from 1662 to 1677. One suspects that he was placed ahead of such men as Boyle and Wallis in the presidency partly on account of his social position, but the Society's records show that he was faithful in attendance and fulfilled the presidential duties conscientiously. The great advantage that came to him from this connexion was that he was fully aware of the great scientific advances then taking place all over the world. He studied mathematics in Oxford, took a degree in medicine, and initiated his scientific activity by a study of the recoil of guns.

Our present business is with Brouncker the mathematician. Here he enjoyed the great advantage of intimacy with Wallis. From this sprang the impetus for most of his scientific activity. In the *Arithmetica Infinitorum*, published in 1665, Wallis undertook to find the approximate value for $4/\pi$. I will not reproduce his long calculation; the explanation in Scott (q.v.) covers pp. 47–60. The final result can be written:

$$\frac{4}{\pi} = \frac{3 \times 3 \times 5 \times 5 \times 7 \times 7 \times \ldots}{2 \times 4 \times 4 \times 6 \times 6 \times 8 \times 8 \times \ldots}. \tag{1}$$

Nowadays we reach this easily from the inequality

$$\int_0^{\frac{1}{2}\pi} \sin^{2n}x \, dx > \int_0^{\frac{1}{2}\pi} \sin^{2n+1}x \, dx > \int_0^{\frac{1}{2}\pi} \sin^{2n+2}x \, dx,$$

or from Stirling's formula

$$\lim_{n \to \infty} \frac{n^n e^{-n} \sqrt{(2\pi n)}}{n!} = 1.$$

Wallis was not satisfied with an answer of this sort. An endless product looked to him suspicious. He might have reasoned that if we write it

$$\frac{9}{8} \times \frac{25}{24} \times \frac{49}{48} \times \ldots,$$

the result is greater than unity, as it should be, whereas if we write it

$$\frac{1}{2}\times\frac{9}{16}\times\frac{25}{36}\times\cdots,$$

the result appears less than $\frac{1}{2}$. In any case he submitted it to Brouncker, who came back with the result

$$\frac{4}{\pi}=1+\cfrac{1}{2+\cfrac{9}{2+\cfrac{25}{2+\cdots}}}\qquad(2)$$

Right here we have a great disappointment, we simply do not know how Brouncker reached this result. There is what seems to be a proof in Wallis[†] which frankly I am unable to follow. He begins by noting that the product (1) can be written, in modern notation,

$$\prod_{r=1}^{r=\infty}\frac{(2r+1)^2}{[(2r+1)^2-1]}.$$

He then considers the function

$$\phi(y)=y+\cfrac{1}{2y+\cfrac{9}{2y+\cfrac{25}{2y+\cdots}}}-$$

and shows that for the first three convergents

$$\phi(y)\phi(y+2)-[\phi(y+1)]^2$$

is small. I do not see how he proceeds from there and agree with Reiff: 'Der Beweis den Wallis dafür giebt, ist aber so gekünstelt, dass man nicht annehmen kann, Brouncker habe denselben Weg eingeschlagen.'[‡]

Here is a very simple proof based on some work of Euler's. Suppose we have a convergent infinite series

$$c_0+\gamma_1+\gamma_1\gamma_2+\gamma_1\gamma_2\gamma_3+\cdots.$$

This is equivalent to the continued fraction

$$c_0+\cfrac{\gamma_1}{1-\cfrac{\gamma_2}{1+\gamma_2-\cfrac{\gamma_3}{1+\gamma_3-\cdots}}}$$

Let us take in particular

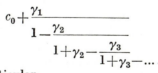

$$\tan^{-1}x=\int_0^x\frac{dt}{1+t^2}=x-\frac{x^3}{3}+\frac{x^5}{5}-\frac{x^7}{7}+\cdots.$$

† q.v., pp. 470 ff.   ‡ Reiff (q.v.), p. 14.

Put $\gamma_1 = x$, $\gamma_2 = -\dfrac{x^2}{3}$, $\gamma_3 = -\dfrac{3x^2}{5},\ldots$

$$\tan^{-1} x = \cfrac{x}{1+\cfrac{x^2}{3-x^2+\cfrac{9x^2}{5-3x^2+\cfrac{25x^2}{7-5x^2+\ldots}}}}$$

Putting $x = 1$

$$\frac{\pi}{4} = \cfrac{1}{1+\cfrac{1}{2+\cfrac{9}{2+\cfrac{25}{2+\ldots}}}} = \cfrac{1}{\dfrac{4}{\pi}}.$$

Perron† gives essentially this, with the comment 'Dieser Kettenbruch war lang vor Euler bekannt'.

At this point we run into an interesting historical question. It has sometimes been assumed that Brouncker and Wallis were the first to make use of continued fractions. This is not the case. Continued fractions were used much earlier in computing square roots.‡ Two Italians, Raffaele Bombelli and Pietro Cataldi, used them for simple root extraction. We find a long account of Cataldi's work in Bortolotti (q.v.). He claims that we owe the latter for

1) The fact that successive convergents are alternately less and greater than the ultimate value.
2) The formulation of the rule for finding convergents in terms of the successive quotients.
3) The estimate of the limit of error.

These are important claims, and I am not sure that Cataldi is entitled to the full credit for the general continued fraction; he does not use perfectly general numerators and denominators, and the only fractions that he uses are those which approach square roots. Independent of these Italians Daniel Schwenter showed how complicated fractions involving large numerators and denominators could be approximated to by continued fractions involving only small numbers. In Schwenter§ we have an approximation to $\frac{233}{177}$. He first carries through the Euclidean process for finding the highest common factor, setting down the quotients and remainders in two columns. In two other columns he sets down the numerators and denominators of the successive convergents, much as we should do it to-day. He evidently had the general idea. I doubt whether either Brouncker or Wallis ever heard of Schwenter,

† q.v., p. 209.   ‡ Tropfke, 2nd ed., vol. vi, pp. 75 ff.
§ q.v., Part II, Book I, Problem XIII, pp. 65 ff.

the number $4/\pi$ in which they were interested is not the root of any equation with rational coefficients. We therefore owe these geometers thanks for being the first to use continued fractions to approximate to a transcendental number. I think it is possible that Wallis surmised the general rule for finding successive convergents; we find him writing:

Esto igitur fractio eiusmodi continue fracta quaelibet sic designata

$$\frac{a}{\alpha}\frac{b}{\beta}\frac{c}{\gamma}\frac{d}{\delta}\frac{e}{\epsilon}\text{ etc.}$$

Cum igitur constet, recepta methodo reductionem institui posse ad hunc modum

$$\frac{a}{\alpha}\frac{b}{\beta}=\frac{a\beta}{b+\alpha\beta}, \qquad \frac{a}{\alpha}\frac{b}{\beta}\frac{c}{\gamma}=\frac{ac+a\beta\gamma}{\alpha c+b\gamma+\alpha\beta\gamma}.$$

Et sic deinceps quantum opus erit.'†

He may, of course, merely have put this through by main force, but again he may have surmised some more general method of calculation.

§ 2. The semi-cubical parabola and the cycloid

Brouncker's next mathematical work appeared in 1659 in Wallis's discussion of the cycloid, but seems to have been completed earlier. The work is interesting because before the methods of Newton and Leibniz were widely known, there were so few curves which could be rectified. Van Schooten had shown that a certain Heurat had rectified the semi-cubical parabola. Wallis wrote to Huygens that this had previously been done by a certain William Neile. Neile's proof was apparently improved on by Wren. Wallis sent Brouncker 'Hanc D. Nelii demonstrationem, ubi conspexerat Illustris Brounckerus suam ille statim quae sequitur non absimilem concinnavit, et impertivit mihi, quam jam ultra duos annos apud me habui'.‡

It is true that the two proofs are much alike, Neile's being, I think, the easier. I am sure, however, that Brouncker was too honourable a man to send in some slight modification, claiming that it was original work. Let us therefore assume that *concinnavit* means that he had already knocked something together and was now sending it along. Here is the essential part.

Let us remember that a semi-cubical parabola is the curve whose ordinates are proportional to the areas of parabolic arches cut by pairs of ordinates perpendicular to the axis. We take the parabolic half-segment ABC

$$AB = a; \qquad BC = b; \qquad BE = c.$$

† Wallis (q.v.), vol. i, p. 475.   ‡ Ibid., vol. i, p. 552.

Let AeE be the semi-cubical parabola, and AcC the corresponding parabola:

$$\frac{\text{Par } ABC}{\text{Par } Aac} = \frac{BE}{ae} = \frac{c}{ae}.$$

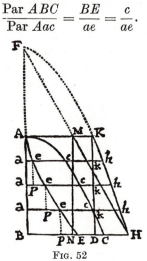

Fig. 52

Brouncker will prove

$$\frac{AB}{AeE} = \frac{27ac^2}{(4a^2+9c^2)[\sqrt{(4a^2+9c^2)}-8a]}.$$

Now c is proportional to Par ABC.

Suppose

$$\frac{AB}{BE} = \frac{AK \times AB}{\text{Par } ABC}.$$

$$\frac{\text{Par } ABC}{AK} = BE; \qquad \frac{\text{Par } Aac}{AK} = ae.$$

Take two values of $\dfrac{\text{Par } Aac}{AK}$ and subtract. Then, approximately,

$$\frac{aa \times ae}{AK} = Pe; \qquad ee^2 = (aa)^2 + (Pe)^2.$$

By the fundamental property of the parabola

$$\frac{ac^2}{b^2} = \frac{Aa}{a};$$

$$ah = \sqrt{(AK^2 + ac^2)}; \qquad BH = \sqrt{(AK^2 + b^2)};$$

$$ah^2 = ac^2 + BH^2 - b^2$$

$$= \frac{b^2}{a} Aa + BH^2 - b^2.$$

Then h traces a parabola where the element of area is

$$aa \times ah = AK \times ee.$$

The problem of finding the length of an arc of a semi-cubical parabola is that of finding the area under an arch of a parabola, solved by Archimedes. It does not seem to me necessary to follow the calculation any farther.

Brouncker's next contribution, Brouncker[2], is called 'A Demonstration of the Synchronism of the Vibrations in a Cycloid' and is very brief, but we find the calculation at length in Birch.[†] The date is 22 January:

'The pendulum experiment was discoursed by Lord Viscount Brouncker who brought in the account and schemes of it, and a committee was appointed for making trials of it, consisting of his lordship himself, Mr. Boyle, Sir William Petty, Dr. Williams, and Dr. Wren.

'His lordship's paper was ordered to be registered, and a copy of it was made against the Friday following and brought to Sir Herbert Moray to be sent to Mons. Huygens.

'The paper was as follows.'

The paper is given in detail. It covers four very large pages; I cannot see that it is in any way an improvement on Huygens. The important question is whether this was the first demonstration of the interesting theorem that the cycloid is the tautochronous curve. Huygens's proof is found in his *Horologium oscillatorium*, which was not written till 1665, but that does not settle the question. The wording above suggests that Huygens had announced the tautochronism of the cycloid, and perhaps sent along his proof, and a committee was appointed to test the thing experimentally. Brouncker then either published a proof which he had worked out previously, or lacking Huygens's proof, worked out one of his own. Cantor inclines to this latter view.[‡] It seems highly likely that Huygens had proved the theorem mathematically, not experimentally, following the lines of his subsequently published proof. This idea is strengthened by the following:

'On examine le mouvement sur les surfaces inclinées, et c'est là qu'on a démontré cette propriété si estimée, que je scay que M. Huygens a démontré touchant le mouvement qui se feroit sur une cycloïde.'[§]

[*]Paradies gives a proof.[||] The date of this writing was subsequent to 1665, but Cantor, loc. cit., points out that it was before the *Horologium oscillatorium* was available. It seems to me that Huygens must have found and communicated to a certain number of people a proof anterior to Brouncker's.

§ 3. The hyperbolic area and logarithms

Brouncker's last mathematical contribution will demand longer consideration, for it involves several other mathematicians. It appeared

† q.v., pp. 70 ff.   ‡ Cantor[1], vol. iii, p. 134.
§ Paradies (q.v.), 16th page of the preface. || pp. 233 ff.
* Paradies; Pardies is the more usual spelling.

in 1667 and was devoted to the much studied computation of the area bounded by an arc of a rectangular hyperbola, the ordinates dropped from its ends on an asymptote, and the intercepted segment of that asymptote. Let us look at the history of this problem.

In 1647 there was published in Antwerp the *Prolegomena a Santo Vincento. Opus geometricum quadraturae circuli et sectionum coni* which is accessible in Bopp[2]. We may assume that Brouncker was familiar with this, for Wallis wrote to to him in 1658:

'Si l'on prend sur l'asymptote les droites *NH, NI, NQ, NK, NL, NM* en proportion géométrique, des points *H, I, Q, K, L, M* on mène des droites parallèles à l'autre asymptote, l'espace hyperbolique *ABHM* est divisé en cinq parties égales, comme l'a démontré Grégoire de St. Vincent Livre X je crois.'[†]

The proof is easy, and comes to this:[‡]

Let the extreme ordinates be $P_0 Q_0$, $P_n Q_n$. We divide the area into equal parts by ordinates $P_1 Q_1$, $P_2 Q_2$,..., $P_{n-1} Q_{n-1}$

$$P_i Q_i \times P_i P_{i+1} = P_{i+1} Q_{i+1} \times P_{i+1} P_{i+2}; \qquad \frac{P_i Q_i}{P_{i+1} Q_{i+1}} = \frac{P_{i+1} P_{i+2}}{P_i P_{i+1}}.$$

But

$$\frac{P_i Q_i}{P_{i+1} Q_{i+1}} = \frac{OP_{i+1}}{OP_i} = \frac{OP_{i+1} + P_{i+1} P_{i+2}}{OP_i + P_i P_{i+1}} = \frac{OP_{i+2}}{OP_{i+1}} = \frac{P_{i+1} Q_{i+1}}{P_{i+2} Q_{i+2}},$$

$$\frac{P_0 Q_0}{P_n Q_n} = \left(\frac{P_{n-1} Q_{n-1}}{P_n Q_n} \right)^n.$$

St. Vincent considers also the incommensurable case, but we need not follow him here. I note that he treats the area as the sum of a number of rectangles; these were the days of Cavalieri's indivisibles.

I turn next to Fermat. I cannot find the date of his publication of the following result, but as he died in 1665 it was before the date of Brouncker's publication of the work with which we are occupied. This theorem, which is in a discussion of quadratures, states that we can find the area bounded in this same way by any hyperbola, save the first, by a uniform method.[§]

His first idea is to take as bases of his infinitesimal rectangles not points equally spaced on the asymptote but those which mark a decreasing geometrical series of distances from a fixed point. Consider the hyperbola $x^n y = 1$ and the points whose abscissae are a, ar, ar^2.... The element of area is then

$$y[ar^m - ar^{m+1}] = \frac{ar^m (1-r)}{a^n r^{mn}} = \frac{1-r}{a^{n-1} r^{m(n-1)}}.$$

[†] French translation of Tannery and Henri, in Fermat (q.v.), vol. iii, p. 576.
[‡] Bopp[2], p. 265.   [§] Fermat (q.v.), vol. iii, pp. 217 ff.

This only breaks down when $n = 1$.

We now pass to James Gregory. In 1667 he laid before the Royal Society a paper entitled *Vera Circuli et Hyperbolae Quadratura*. Here is the comment:

'This treatment perused by some very able and judicious mathematicians, and particularly by the Lord Viscount Brouncker, and the Reverend John Wallis, received the character of being very ingeniously and mathematically written, and well worth the study of men addicted to that science; that in it the author has delivered a new analytical method for giving the aggregate of an infinite or indefinite converging series, and that thence he teaches a method of squaring the circle, ellipsis and hyperbola by infinite series, and calculating the true dimensions as close as you please. And lastly that by that same method from the hyperbola he calculates both the logarithm of any number assigned and, vice versa, the natural number of any logarithm assigned.'[†]

This seems to be the first statement of the connexion of logarithms with the hyperbola. Let us look into the question of who first saw this connexion. Cantor frankly gives the credit to St. Vincent, basing his statement on the theorem proved on p. 142, although St. Vincent never uses the word logarithm.

'Das ist offenbar die Wahrheit von der Quadratur der auf die Asymptoten bezogenen Hyperbel durch Logarithmen, welche hier entdeckt ist, ohne dass Gregorius sich dabei des Wortes Logarithmen bedient habe.'[‡]

A much more definite statement is found in Hutton:

'As to the first remarks on the analogy between logarithms and hyperbolic space, it having been shown by Gregory St. Vincent in his *Quadratura Circuli et Sectionum Coni* published at Antwerp in 1647, that if an asymptote be divided into parts in geometrical progression, and if from the points of division ordinates be drawn parallel to the other asymptote, they will divide the space between the asymptote and the curve in equal parts, from thence it was shown by Mersenne that by taking the continual sums of these parts there would be obtained areas in arithmetical progression which therefore were analogous to a system of logarithms. And the same analogy was remarked and illustrated soon after by Huygens and others.'[§]

This statement is important, if correct, but Hutton does not rank as a perfectly accurate historian. I find no confirmation in any of the standard histories of mathematics, except that already quoted from Cantor. Montucla is very much interested in the point:

'Cette propriété est du plus grand usage dans la géométrie transcendante et elle a fourni la résolution pratique de tous les problèmes qui dépendent de la quadrature d'un espace hyperbolique à l'usage d'une table de logarithmes. Au reste la découverte de cette propriété est revendiquée par plusieurs géomètres.'[||]

† Gregory (q.v.), p. 232.   ‡ Cantor[1], vol. ii, p. 896.
§ Hutton (q.v.), p. 86. || Montucla (q.v.), vol. ii, p. 80.

We find in Kästner† an attack on St. Vincent by Mersenne, who blames him for the fact that he does not show how to solve the problem, given three positive numbers, and the logarithms of two of them, to find the logarithm of the third. Kästner goes on to give the answer of Sarsa, who says the problem is easy when the numbers are commensurable. Montucla points out that the logarithm is there in every case. If we have given a straight line and two points on one side of it, there is just one rectangular hyperbola through these two points which has the line as an asymptote. I cannot find the passage in any of Mersenne's works accessible to me. It is certainly no great step to pass over from St. Vincent's series theorem; such a step would seem quite natural to anyone who considered the question. For instance, we find Mercator dealing with the hyperbola:

$$y = \frac{1}{x+1} = 1-x+x^2-x^3+\dots$$

$$\int y \, dx = \log(1+x) = x - \frac{x^2}{2} + \frac{x^3}{3} - \frac{x^4}{4} + \dots.$$

Of course he did not bother with any question of convergence.‡

It is high time to return to Lord Brouncker. On the page of the *Philosophical Transactions* facing that with the mention quoted above of Gregory's work, we find Brouncker on *The Squaring of the Hyperbola*.§ I showed on p. 142 how Fermat had the idea of spacing ordinates from an asymptote from lengths which follow a geometrical progression. Brouncker's scheme is rather more complicated. He starts with Mercator's hyperbola:

$$(x+1)y = 1.$$

He wishes to find the area bounded by the X-axis, which is an asymptote, the ordinates $x = 0$, $x = 1$, and the intervening arc of the curve. We first go to the point $(1, \frac{1}{2})$ and draw the coordinates. We have with the axes a rectangle whose area is $1/(1 \times 2)$. We then go to the point $(\frac{1}{2}, \frac{2}{3})$ and draw down an ordinate as far as the top of the preceding rectangle and an abscissa as far as the Y-axis. We add to our area a rectangle of

$$\frac{1}{2}\left(\frac{2}{3} - \frac{1}{2}\right) = \frac{1}{3 \times 4}.$$

We then go to the points $(\frac{1}{4}, \frac{4}{5})$, $(\frac{3}{4}, \frac{4}{7})$ and form rectangles on the previous model; their areas will be

$$\frac{1}{5 \times 6}, \qquad \frac{1}{7 \times 8}.$$

† q.v., vol. iii, p. 251.   ‡ Zeuthen (q.v.), p. 314.
§ *Phil. Trans.*, abridged ed., 1809, vol. i, p. 233.

Here is the general scheme, although Brouncker does not give the details. We go to the point

$$\left(\frac{2m+1}{2^n}, \frac{2^n}{2^n+2m+1}\right)$$

and drop perpendiculars on the lines

$$x = \frac{2m}{2^n}, \qquad y = \frac{2^n}{2^n+2m+2}.$$

We form a rectangle whose area is

$$\frac{1}{(2^n+2m+1)(2^n+2m+2)}.$$

For each n the limits for m are 0, 2^{n-2}. We reach finally

$$\frac{1}{1\times2}+\frac{1}{3\times4}+\frac{1}{5\times6}+\frac{1}{7\times8}+\cdots.$$

But Brouncker does not stop here, he calculates similarly the area on the other side of the curve, that is to say, bounded by the curve and the lines $x = 1$, $y = 1$. Starting with the same point

$$\left(\frac{2m+1}{2^n}, \frac{2^n}{2^n+2m+1}\right)$$

we drop perpendiculars on the lines

$$x = \frac{2m+2}{2^n}, \qquad y = \frac{2^n}{2^n+2m}.$$

We thus get the area

$$\frac{1}{(2^n+2m)}\times\frac{1}{(2^n+2m+1)},$$

and so the series

$$\frac{1}{2\times3}+\frac{1}{4\times5}+\frac{1}{6\times7}+\cdots.$$

I repeat that Brouncker does not give this general formula, merely taking it for granted that the procedure which worked in the first three cases can be stepped up indefinitely.

The noble viscount then takes the interesting step of calculating another series which converges more rapidly, namely that which gives the area between the curve and the chord connecting the end points $(0, 1)$, $(1, \frac{1}{2})$. This he gets from an infinite series of triangles, much as Archimedes did in calculating the area of a parabolic segment. We have triangles with vertices

$$\left(\frac{2}{2^n}, \frac{2^n}{2^n+2m}\right), \qquad \left(\frac{2m+1}{2^n}, \frac{2^n}{2^n+2m+1}\right), \qquad \left(\frac{2m+2}{2^n}, \frac{2^n}{2^n+2m+2}\right).$$

Instead of using these coordinates and the formula for the area of their triangle, he makes use of the simple theorem that if a point be within a rectangle, the area of the triangle whose vertices are this point and the ends of a diagonal is one-half the difference between the areas which the lines through the point parallel to the sides make with the sides across the diagonal and with the sides on the same side of the diagonal. This gives him the form he seeks:

$$\frac{1}{2}\left(\frac{1}{2\times3}-\frac{1}{3\times4}+\frac{1}{4\times5}-\frac{1}{5\times6}+\ldots\right)=\frac{1}{2\times3\times4}+\frac{1}{4\times5\times6}+\frac{1}{6\times7\times8}+\ldots.$$

At this point Brouncker becomes interested in convergence. He remarks:

$$\frac{1}{1\times2}+\frac{1}{2\times3}+\frac{1}{3\times4}\ldots=1,$$

not for the fairly obvious reason that the sum of the areas on the two sides of the curve is unity, but because if a be the number of terms taken at pleasure, the last term is $1/(a^2+a)$ and the sum so far $1-1/(a+1)$. He does not give a complete proof based on mathematical induction. He then tries with less success to show that the difference between the sum of his third series and the area sought is less than any assigned quantity. He does show that if in any area of this sort the end ordinates are commensurable, the area can be found as closely as we wish by this device.† It seems to me interesting that at this early period in the history of infinite series Brouncker not only bothered with questions of convergence, but also of showing that a series converged to the right value.

What should be our final judgement of Brouncker as a mathematician? I am afraid that I cannot be as enthusiastic as I should like to be. One must be more exacting in judging a man who had every scientific advantage that his age offered, and was in touch with all the ablest mathematicians of his time, than in the case of a man less favourably placed. He was certainly an able man, and his continued fraction approximation to $4/\pi$ was admirable. But in all of the other papers I have mentioned he was pretty close to what others had done, and sometimes done better; a sad verdict.

† Zeuthen (q.v.), pp. 310–13.

CHAPTER XII

GUILLAUME L'HOSPITAL, MARQUIS DE SAINTE-MESME

§ 1. Original contributions

GUILLAUME FRANÇOIS L'HOSPITAL, Marquis de Sainte-Mesme, Comte d'Entremont, Seigneur d'Ouques, etc., was born in 1661 and died in 1704. He was a shining example of a man of the highest social distinction whose love of learning drove him to devote much of his short life to scientific writing. There were, as we shall see, certain inexactitudes in his mathematics, but they were nothing compared to the vagaries in the spelling of his name: 'He is also known as the Marquis de l'Hospital. The family also spelled the name Lhospital and, somewhat later l'Hôpital.'†

A young man sprung from a socially important family was naturally destined to serve in the armed forces, and there Guillaume began, rising to the rank of captain, but withdrew from the army because his extreme near-sightedness proved an insurmountable barrier. It is interesting to note in this connexion that Taine tells us in his *Histoire du Consulat et de l'Empire* that the reason why the French won the great victory of Auerstädt was that Marshal Davoust was very near-sighted, and always went way forward to get a good view of the enemy; this enabled him to discover that a large German force was endeavouring to withdraw after the great defeat at Jena. One is inclined to believe that it was L'Hospital's love of mathematics rather than the imperfection of his vision that led him to abandon a military career in favour of a scientific one. He early established contact with Huygens, Leibniz, James and John Bernoulli, and other scientists. A most significant event occurred in 1691 when John Bernoulli came to spend a season in the French capital, and passed part of his time at L'Hospital's estate of Ouques. This was significant in the lives of both. L'Hospital subsequently secured for the Swiss a position as professor at the University of Groningen.‡ This was in return for being inducted into the mysteries of Leibniz's new doctrine of differences, and receiving a start towards mathematical production. Their subsequent relations were not always entirely happy, as we shall see.

L'Hospital's first contribution to mathematics was his 'Solution du problème que M. de Beaune proposa autrefois à M. Descartes', which appeared in the Dutch *Journal des Sçavans* for September 1692. Right here we run into a priority dispute. Descartes was unable to solve the problem; Leibniz apparently solved it, but did not publish his solution.

† Smith (q.v.), vol. i, p. 384.　　　　　　　‡ Rebel (q.v.), p. 13.

Bernoulli and L'Hospital seem to have studied it together, which subsequently led to heart-burning.† Here is the problem: 'To find a curve through the origin which has the property that the slope of the tangent at any point is equal to a fixed quantity N divided by the length on the ordinate between the point of contact, and the intersection with the line $x = y$.' L'Hospital does not set up and solve the differential equation, but gives a geometrical construction that produces the result

$$\frac{dy}{dx} = \frac{N}{y-x}, \qquad \frac{dy}{N} + \frac{d[N-(y-x)]}{N-(y-x)} = 0,$$

$$\frac{y}{N} + \log[N-(y-x)] = 0. \tag{1}$$

On p. 596 of the *Journal des Sçavans* he proceeds as follows. He takes the hyperbola
$$(N-\xi)(N-\eta) = N^2.$$

He then finds the area bounded by the lines $x = 0$, $y = \bar{y}$ and the arc of the curve from $(0, 0)$ to $(\bar{x}, \bar{y})$. In his picture $\bar{x} < 0$, $\bar{y} > 0$ and the curve sought is in the north-east quadrant. This area he calls Nx.

$$Nx = -\int_0^{\bar{y}} \xi \, d\eta = -\int_0^{\bar{y}} \left(N + \frac{N^2}{\eta - N}\right) d\eta$$

$$= -N\bar{y} + N^2\log(N-\bar{y}).$$

We then put $y - x = \bar{y}$ and fall back on (1).

A similar but more difficult problem appeared a year later in the *Mémoires de l'Académie des Sciences*, vol. x, June 1693. We seek a curve which has the property that the length on the tangent measured from the point of contact to the intersection with the X-axis bears a fixed ratio to the distance from the origin to the intersection of that axis with the tangent. Here again it is easy to set up the differential equation

$$\frac{y\sqrt{(dx^2+dy^2)}}{x\,dy - y\,dx} = \frac{p}{q}. \tag{2}$$

L'Hospital does not tell us how to integrate this equation, he merely states: 'Les principes sont peu connus.' He gives a solution involving an auxiliary variable z‡:

$$\frac{x}{y} = \frac{z^2 + (q^2 - p^2)a^2}{2paz}, \qquad y = \frac{a^{2+p/q}z^{1-p/q}}{z^2 + (p+q)^2a^2}. \tag{3}$$

This curve gives a solution of the differential equation. It is rather curious that p and q appear explicitly, not merely through their ratio,

† Rebel (q.v.), pp. 3, 4.   ‡ Ibid., p. 28.

which is the way in which they appeared at first. To find a geometrical construction he takes two rectangular hyperbolas,

$$\xi y = q - p; \qquad z\eta = -q.$$

We get the relation of x and y to z by comparing areas under these curves.

In the tenth volume of the *Mémoires de l'Académie* L'Hospital has two short notes. He begins by seeking the point where a caustic by refraction touches its envelope. This problem had already been solved by others, and I shall return to it again. On pp. 397–8 of the same journal he raises a more interesting point. He says that Leibniz, Huygens, and others insisted that at a point of inflexion the radius of curvature is infinite whereas, says L'Hospital, it may be 0. He gives as an example the curve

$$x^3 = y^5,$$

which appears to the eye to have an inflexion at the origin, though the radius of curvature vanishes. The difficulty here is a lack of a clear definition for an inflexion. This is usually treated as a non-singular point of the curve, but as every line through the origin will have at least three coincident intersections with the curve, this point is singular. L'Hospital naturally knew nothing of all this, his point is well taken. Unfortunately he makes another mistake at this point. He says that if a curve have an inflexion at any point, that involute which passes through the point will also have an inflexion there. This is not the case analytically.

The next article of L'Hospital's which I care to mention appeared in the *Acta Eruditorum* for February 1695. A drawbridge is hoisted by a rope passing over a pulley to which is attached a weight sliding down a given curve. What should be the form of this curve if the system is to be in equilibrium at every point? L'Hospital makes two modifying assumptions: the rope is attached to the drawbridge at its centre of gravity, and the pulley is just above the hinge. As a matter of fact he makes the height of the pulley equal to the weight of the bridge, that is merely a question of units. It simplifies the result, for he remarks *ex statica* the tension in the rope is then equal to the distance from the pulley to the point of attachment to the bridge, or from the weight down along the line of the rope produced to the intersection with a circle whose centre is the pulley, and whose radius is the length of the rope. If the length of the rope be a and the weight b, we find by similar triangles

$$\left\{ \frac{ay}{\sqrt{(x^2+y^2)}} - y \right\} : \left\{ b+x - \frac{ax}{\sqrt{(x^2+y^2)}} \right\} = dx : dy,$$

$$(b+x)\,dx + y\,dy = a d\sqrt{(x^2+y^2)},$$

$$bx + \tfrac{1}{2}(x^2+y^2) = a\sqrt{(x^2+y^2)} + K.$$

This note is followed by two from the Bernoulli brothers. John shows that in the general case we have a trochoid, of some sort; his brother James carries the thing through very easily, removing the restrictions placed by L'Hospital.

In the July number of the same periodical we find another pretty

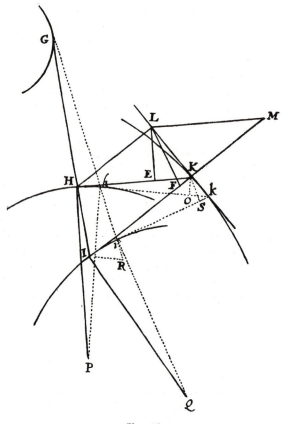

Fig. 53

problem of L'Hospital's. A straight line moves so as always to be tangent to a given curve, and to cut two others. The tangents are drawn at the two points of cutting, and brought to intersect. To draw a tangent to the locus of this meeting-point. This problem is most easily solved by line geometry. Here is L'Hospital's solution by means of the elementary geometry of infinitesimals.

Let H, I be the cutting points, while G is the point of contact with the envelope of the line. K and k are nearby points on the locus sought.

Let $GH = a$, $GI = b$, $HK = m$, $IK = n$.

The radii of curvature at H and I shall be f and g.

Let α_1 and α_2 be the angles which the tangents HK, IK make with some fixed direction. If L be any point on the tangent to the locus of k, the ratio of its distances to the two lines HK, IK is the ratio for k.

$$\frac{LE}{LF} = \frac{KO}{KS} = \frac{m\Delta\alpha_1}{n\Delta\alpha_2} = \frac{m\dfrac{\Delta\alpha_1}{\Delta t_1}Hh}{n\dfrac{\Delta\alpha_2}{\Delta t_2}Ii} = \frac{m}{n}\times\frac{g}{f}\times\frac{a}{b}\times\frac{Ih}{Ii} = \frac{am^2g}{bn^2f}.$$

A pretty piece of geometry.

In the *Acta Eruditorum* for May 1697 is a long discussion of the problem of the brachistochrone, or curve of quickest descent under gravity. L'Hospital is sometimes given credit for being one of the first to solve this, but this article is very short, so that unless he has made some other publication in another place which I have not seen, I cannot see that he deserves much credit. He does, however, about this time, solve some rather neat problems. Here is one from the January 1698 number of the same journal.

A series of ellipses have a common major axis. How shall we find the tangent to the curve which is the locus of such a point that the area bounded by the ordinate, the arc to the major axis, and the segment of that axis back to the foot of the ordinate is constant? We must remember that such an area bears a simple relation to the corresponding arc for the major auxiliary circle.

Let the moving point be (x, y) on the variable ellipse

$$\frac{x^2}{a^2} + \frac{y^2}{t^2} = 1.$$

The corresponding point on the major auxiliary circle will be $(x, ay/t)$

$$t\left[a^2\cos^{-1}\frac{x}{a} - x\sqrt{(a^2-x^2)}\right] = c,$$

$$\frac{x^2}{a^2} + \frac{y^2\left[a^2\cos^{-1}\dfrac{x}{a} - x\sqrt{(a^2-x^2)}\right]^2}{c^2} = 1.$$

By definition,

$$d\left[a^2\cos^{-1}\frac{x}{a} - x\sqrt{(a^2-x^2)}\right] = ay\frac{dx}{t},$$

$$\frac{x}{a^2}dx + \frac{y}{ct}\left[\frac{ay^2}{t}dx + \frac{cy}{t}dy\right] = 0.$$

Putting $\dfrac{y^2}{t^2} = \dfrac{a^2-x^2}{a^2}$ we have the differential equation; L'Hospital omits the analysis.

In the August 1699 number of the *Acta* L'Hospital gives a *Solutio facilis et expedita methodus inveniendi solidi rotundi*. This is the problem of finding which form of tube will offer the least resistance to a liquid flowing through it in the direction of the axis. This had already been treated by Newton. L'Hospital finds his treatment too long, and offers something simpler. It certainly is simpler, but it seems to me that he oversimplifies the physical assumptions, and does not see nearly as far into the problem as had the Englishman.

In the *Mémoires de l'Académie des Sciences* for 1700 we find another pretty geometrical problem in an article entitled 'Solution d'un problème physico-mathématique'. This is to find the shape of a curve such that a body descending it under its own weight will always exert a normal pressure equal to that weight. We start with Huygens's principle that if a body move down under gravity along a frictionless path in an upright plane, the normal acceleration is equal to the square of the velocity divided by the radius of curvature. Let us assume that the mass is unity, the arc length s, and this we take as the independent variable,

$$ds = \sqrt{(dx^2+dy^2)}; \qquad d^2s = 0; \qquad dxd^2x+dyd^2y = 0;$$

$$d^2y = -\frac{dx}{dy}d^2x.$$

For the radius of curvature

$$\frac{1}{z} = \frac{dyd^2x - dxd^2y}{ds^3} = \frac{d^2x}{dyds}. \tag{4}$$

The normal force exerted by a unit mass will be

$$-\frac{1}{g} \cdot \frac{d^2x}{dyds}\left(\frac{ds}{dx}\right)^2 + \frac{dx}{ds} = 1.$$

Now by the elementary principles of mechanics, for a body coming down a curve in a vertical plane

$$-\frac{d^2s}{dt^2} = g\frac{dy}{ds}; \qquad -\frac{ds}{dt}\cdot\frac{d^2s}{dt^2} = g\frac{dy}{dt}; \qquad -\frac{1}{g}\left(\frac{ds}{dt}\right)^2 = 2y.$$

From (4), $\dfrac{2y\,d^2x+dydx}{2\sqrt{y}} = \dfrac{dsdy}{2\sqrt{y}}.$

This he integrates ingeniously:

$$d[dx\sqrt{y}] = d[ds\sqrt{y}];$$

since $d^2s = 0; \qquad dx\sqrt{y} = ds[\sqrt{y}-\sqrt{a}];$

since ds is constant. Replacing ds by its value,

$$dx = \frac{dy(\sqrt{y}-\sqrt{a})}{\sqrt{\{2\sqrt{(ay)}-a\}}}.$$

It is a perfectly straightforward job to integrate this, giving, surprisingly enough, an algebraic curve. The whole seems to me his
prettiest piece of mathematical work.

L'Hospital's last original paper appeared in the *Mémoires de l'Académie*

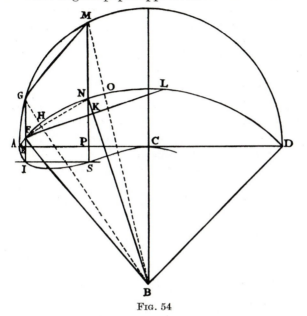

Fɪɢ. 54

des Sciences for 1701. It is entitled 'La quadrature absolue d'une
infinité de portions moyennes'. Once more we are dealing with figures
bounded by circular arcs, which we have seen were so dear to the heart of
Leonardo da Vinci. Again we meet the theorem that if the radii of two
circles are in the ratio $\sqrt{2}:1$ a sector of the first with a central angle of θ
has the same area as a sector of the second with an angle of 2θ. L'Hospital's problem is like a lot solved by Leonardo, merely a bit more general
and difficult. He refers to Hippocrates, but not to the Italian geometer.

We have two circles as in Fig. 54. We draw a curve ASC according
to the principle $PS = \sin \angle NBM$. Through S we draw a horizontal
line meeting the same curve again in I and through this another
vertical. L'Hospital seeks the area $FGMN$, two sides being circular arcs.

By definition arc $FL = 2\,\text{arc}\,FN = 2\,\text{arc}\,HO = \sqrt{2}\,\text{arc}\,GM$.

Then arc FL and arc GM correspond to the same central angle

$$\text{segment}\ GM = \tfrac{1}{2}\,\text{segment}\ FL$$
$$= \text{segment}\ FN + \tfrac{1}{2}\Delta FNL,$$
$$FGMN = \text{trapezoid}\ FGMN + \text{segment}\ GM - \text{segment}\ FN$$
$$= \text{trapezoid}\ FGMN + \tfrac{1}{2}\Delta FNL.$$

This is L'Hospital's theorem; there are an infinite number of these rectifiable figures, since one vertex can be taken at random on the appropriate arc.

I think we cannot escape the conclusion that as an original mathematician L'Hospital was very much an amateur. He considered mostly problems which had been set and solved by others. He was not above making mistakes, and we have from him no striking contribution to mathematical theory.

§ 2. *L'Analyse des infiniment petits*

After giving such a tepid account of L'Hospital as an original mathematician, it is a pleasure to turn to what he accomplished as a text-book writer. Here he was supreme, a worthy member of the great French school which included Monge and Lacroix.

L'Hospital's first book is his *Analyse des infiniment petits*, first published in 1696. It is an introduction to the differential calculus as conceived by Leibniz, and expounded with all of the enthusiasm of a neophyte. He conceives that there exist in nature magnitudes of such different sizes that we may look upon some as infinitely small compared with others. The small may then be discarded when compared with the large ones. The distinction between this and our modern point of view is that a differential, or, as he says a difference, is not a variable approaching a limit under defined circumstances, but an actual small quantity. The difference of the sum of two finite quantities is the sum of their differences, so he has his first equation

$$d(x+y-z) = (x+dx+y+dy-z-dz)-(x+y-z) = dx+dy-dz.$$

The difference of a product is found equally easily. As for the difference of a quotient, the existence of this is first assumed, then it is found from the product formula

$$dy = d\left[x . \frac{y}{x}\right] = xd\left(\frac{y}{x}\right)+\frac{y}{x}dx.$$

I will mention at this point that L'Hospital never explicitly differentiates the logarithm, although he knows the value of $\int dx/x$. He also does not differentiate the trigonometric functions, although it seems likely that their derivatives were known in his time. This frequently involves him in needlessly long calculations. The differentiation of algebraic expressions goes through very smoothly. There is no integration; it is possible that he planned at one time to write separately on this subject.

In section II we are concerned with tangents to curves. He does not look for the equation of the tangent, but the value of the subtangent, that is to say the orthogonal projection on the axis, usually the X-axis, of as much of the tangent as runs from the point of contact to that axis. This is, of course, $(y\,dx)/(dy)$.

Here are some interesting examples:

$$y^3 - x^3 = axy,$$

$$\frac{y\,dx}{dy} = \frac{3y^3 - axy}{3x^2 + ay}.$$

The distance along the axis from the origin to the tangent is

$$x - \frac{y\,dx}{dy} = \frac{-axy}{3x^2 + ay} = -t,$$

$$y = \frac{3tx^2}{a(x-t)}.$$

Assuming that t does not become infinite with x and y, we put $y = 3tx/a$

$$(27t^3 - a^3)x^3 = 3a^3tx^2,$$

$$x = \frac{3a^3t}{27t^3 - a^3}.$$

When x is infinite, $3t = a,$ $\dfrac{y}{x} = \dfrac{3t}{a} = 1$.

The asymptote makes an angle of 45° with the axis; strange mathematics! In proposition X we find a curious theorem which he ascribes to Tschirnhausen. Given the curve $f(x, y) = 0$ and the point (x_1, y_1), connect $(x+dx, y+dy)$ with this and drop a perpendicular on this line from (x, y). The distance to the point $(x+dx, y+dy)$ from the foot of this perpendicular is, to the first order of infinitesimals,

$$d\sqrt{\{(x-x_1)^2 + (y-y_1)^2\}} = \frac{(x-x_1)\,dx + (y-y_1)\,dy}{\sqrt{\{(x-x_1)^2 + (y-y_1)^2\}}}.$$

Now take a point at a distance r from (x, y) on the line from there to (x_1, y_1)

$$\xi = x + \frac{r(x-x_1)}{\sqrt{\{(x-x_1)^2 + (y-y_1)^2\}}}; \qquad \eta = y + \frac{r(y-y_1)}{\sqrt{\{(x-x_1)^2 + (y-y_1)^2\}}}.$$

The distance from here to the normal at (x, y) is

$$\frac{r\{(x-x_1)\,dx + (y-y_1)\,dy\}}{ds\sqrt{\{(x-x_1)^2 + (y-y_1)^2\}}}.$$

The ratio of this to $d\sqrt{\{(x-x_1)^2 + (y-y_1)^2\}}$ is a function of r alone.

Section III deals with maxima and minima. When a function is represented by a curve with a continuously turning tangent, a maximum or minimum, when not at the end of an interval (a possibility he does not consider) will either be at the top or bottom of an arch, or at a cusp.

On p. 49 he gives Huygens's solution of Leonardo's brilliant point problem, which we saw on p. 54. He does not make it quite clear why this should be a problem in maxima and minima. Next comes the classical refraction problem, the quickest path between two media of different densities. The section closes with an amusing problem taken direct from John Bernoulli's *Differentialrechnung*: To find on what day in the year twilight will be the longest at a point of given latitude. Twilight is supposed to last until the sun has crossed a circle parallel to the horizon, and a certain number of degrees below it. Our variable is the sun's declination; we wish to maximize the difference in the sun's hour angle between the time of geometrical sunset and the crossing of this circle.

In section IV he comes to grips with higher derivatives, or second differences as he calls them. I should mention at this point that L'Hospital is very careful in his choice of independent variable, whose second difference is 0, a care which Pascal also took, as we saw on p. 96. The first application is to points of inflexion. A point of inflexion is, for him, a point where the curvature reverses while the moving point does not; at a cusp the tracing point goes back in its tracks. He does not consider a beak where the point reverses and the curvature does not.

The distance from the origin to the intersection of the X-axis with the tangent is

$$t = \frac{y\,dx}{dy} - x; \qquad dt = \frac{-y\,dx d^2 y}{dy^2}; \qquad d^2 x = 0.$$

In case of an inflexion t will have an extreme value dt, will change from positive to negative or vice versa. He naturally assumes that it will pass through the value 0. In the case of a cusp dt does not change sign whereas dx does. $d^2 y$ must change sign; he assumes that it becomes infinite. The absurdity of assuming that a second difference, which is negligible compared with a first difference, should become infinite should have shown him that there was something fundamentally wrong in his assumptions. Had he realized that in kinematical terms a cusp was a point of 0 velocity, he would have seen that both dx/dt, dy/dt change sign and both $d^2 x/dt^2$, $d^2 y/dt^2$ vanish. He hunts also for inflexions when the *appliqués* are radiating lines. Here we are dealing essentially with polar coordinates which were invented by James Bernoulli in 1691, but did not come into general use for another century. He writes y where we should write r, and dx where we should write $r\,d\theta$. Here is an example where he finds a cusp. This is the curve we saw on p. 148.

$$\frac{y\sqrt{(dx^2 + dy^2)}}{x\,dy - y\,dx} = \frac{p}{q},$$

$$d^2 y = \frac{p(dx dy - x\,dy^2)\sqrt{(dx^2 + dy^2)}}{qy^2\,dy - pxy\sqrt{(dx^2 + dy^2)}}.$$

He takes the two cases $d^2y = 0$; $1/d^2y = 0$. Unfortunately he makes a mistake in each.

In section V we come to evolutes, beginning, however, with involutes. Following Huygens's classical discussion he shows that if a thread be unwound from a given curve, the locus of a point fixed in the thread is an orthogonal trajectory of the system of tangents to the curve. On the other hand, to find the centre of curvature of a given curve, we find where a normal meets another infinitely near, or a point whose immediate motion is down the normal. We can abbreviate his work as follows:

Let the tangent make an angle θ with the horizontal.

$$\frac{dx}{ds} = \cos\theta; \qquad \frac{dy}{ds} = \sin\theta;$$

$$\xi = x - r\sin\theta; \qquad \eta = y + r\cos\theta;$$

$$d\xi = dx - dr\sin\theta - r\cos\theta\,d\theta; \qquad d\eta = dy + dr\cos\theta - r\sin\theta\,d\theta.$$

If
$$dxd\xi + dyd\eta = 0;$$

$$0 = ds^2 - r\,dsd\theta; \qquad \frac{1}{r} = \frac{d\theta}{ds}. \tag{5}$$

There are various forms for this. If we take s as the independent variable $d^2s = 0$ and we have the previous equation:

$$\frac{1}{r} = \frac{dyd^2x - dxd^2y}{dy^2} = \frac{d^2x}{dyds}. \tag{6}$$

If
$$d^2x = 0, \qquad \frac{1}{r} = \frac{dxd^2y}{ds^2}. \tag{7}$$

L'Hospital likes the horizontal projection

$$z = r\cos\theta = \frac{ds^2}{d^2y}. \tag{8}$$

There follow certain well-chosen examples as the conic sections, the *courbe logarithmique ordinaire* $dy = (r\,dx)/a$, and the logarithmic spiral where in our notation $dr = ar\,d\theta$. The interesting thing here is that he must have known how to differentiate the logarithm. He ends with rolling curves, very well done.

In sections VI and VII we have an extended discussion of caustics. A caustic is the envelope of lines, issuing from a given point, which are reflected or refracted at a given curve. The first problem is to find where such a ray will touch its envelope. The usual method of setting up the differential equation is pure geometry in the infinitesimal domain; we can save much labour by a slight use of trigonometry, which L'Hospital, for some strange reason, avoids. Let us begin with the caustic by reflection. Let the source of light be O, the point where it meets the curve P, the centre of curvature C, the orthogonal projection of this point on the reflected ray produced G, while the point where the reflected ray touches its envelope is F. Let OP make an angle ϕ with the direction of the X-axis, the normal make an angle θ, and the reflected ray an angle ψ.

$$\psi - \theta = \theta - \phi,$$

$$PF \, d\psi = dx \sin\psi - dy \cos\psi = \cos(\psi - \theta) \, ds,$$

$$PC \, d\theta = dx \sin\theta - dy \cos\theta = ds,$$

$$OP \, d\phi = -PC \, d\theta \cos(\psi - \theta)$$

$$= -PG \, d\theta,$$

$$PF = \frac{PG \times OP}{2OP + PG}. \tag{9}$$

When the light is at an infinite distance

$$PF = \frac{PG}{2}.$$

When it comes to finding a caustic by refraction we have essentially the same equations as before except that the first is replaced by

$$\frac{\sin(\theta - \phi)}{\sin(\psi - \theta)} = -\frac{m}{n}.$$

In section VIII L'Hospital discusses the envelopes of systems of curves. He takes it for granted that the point where a curve meets an infinitely near member of the family is the point where it touches the envelope, the curve to which all members are tangent. This is capable of proof, but no proof is given. His general procedure is like this. We start with a curve

$$f(x, y) = 0$$

and a variable curve, solving simultaneously for ξ and η

$$F(x, y, \xi, \eta) = F(x + dx, y + dy, \xi, \eta) = 0.$$

We know the ratio of dy/dx from the first equation, and so we can find $\xi - x$ and $\eta - y$ in terms of known quantities. He gives his answer in terms of such things as sub-tangents and sub-normals. In modern times we should write

$$\frac{dy}{dx} = -\frac{f_x}{f_y} = -\frac{F_x}{F_y}.$$

Newton knew this method of writing the slope of a tangent in terms of partial derivatives, but I do not find it explicitly in L'Hospital. All will be clearer if I give one or two examples:

Problem] *Two curves are so related that the ordinate of the first is equal to the normal to the second from the foot of that ordinate. How are their equations related?*

L'Hospital shrewdly remarks that the second curve is the envelope of circles whose centres are the feet of the ordinates to the first curve, and whose radii are equal to those ordinates. We have, then,

$$f(x, y) = 0, \qquad (\xi - x)^2 + (\eta - y)^2 = y^2,$$

$$\xi - x = -y\frac{dy}{dx}.$$

Problem] *Given the curve*

$$f(x, y) = 0,$$

where will the line $\dfrac{X}{x} + \dfrac{Y}{y} = 1$ *touch its envelope?*

$$Xy + Yx = xy; \qquad X\,dy + Y\,dx = x\,dy + y\,dx; \qquad X(x\,dy - y\,dx) = x^2\,dy.$$

The section ends with this question: A point P moves on a certain curve, PQ and PR are the tangents to two other curves. Where will QR touch its envelope? The treatment is rather clumsy; he notes that an easier discussion is based on what we saw on p. 151.

The ninth section is given to the 'solution de quelques problèmes qui dépendent des méthodes précédentes'. The first is the classical problem of indeterminate forms, which he writes:

$$f(a) = \phi(a) = 0, \qquad y = \frac{f(x)}{\phi(x)}.$$

He then puts
$$y = \frac{f(x + dx)}{\phi(x + dx)} = \frac{df}{d\phi}.$$

For instance,
$$\frac{\sqrt{(2a^3x - x^4)} - a\sqrt[3]{(a^2x)}}{a - \sqrt[4]{(ax^3)}}.$$

Differentiating,
$$\frac{\dfrac{a^3 - 2x^3}{\sqrt{(2a^3x - x^4)}} - \dfrac{a\sqrt[3]{a^2}}{3\sqrt[3]{x^2}}}{\dfrac{-3\sqrt[4]{a}}{4\sqrt[4]{x}}}.$$

The limit as $x \to a$ is $16a/9$.

As a matter of fact this particular problem had worried him a good deal. We find him writing in July 1693 to John Bernoulli suggesting that we should substitute directly in the original equation, getting

$$\frac{a^2 - a^2}{a - a} = 2a,$$

and in September of the same year he writes:

'Je vous avoue que je me suis fort appliqué à résoudre l'équation

$$\frac{\sqrt{(2a^3x-x^4)}-a\sqrt[3]{(a^2x)}}{a-\sqrt[4]{(ax^3)}} = y$$

lorsque $x = a$, car ne voyant point de jour pour y réussir puisque toutes les solutions qui se présentent d'abord ne sont pas exactes.'[†]

All this suggests that L'Hospital learnt the correct solution from Bernoulli, but did not give him the specific credit, with the unfortunate result that the method came to be known as L'Hospital's method.

The tenth and final section is given to a 'nouvelle manière de se servir du calcul des différences dans les courbes géométriques, d'où l'on déduit la Méthode de Mrs Descartes et Hudde'.

Here at last we find a small amount of partial differentiation. A maximum or a minimum, not at the end of an interval, will come at the top or the bottom of an arch, or at a cusp. It is clear that at the top or bottom of an arch a slight change in x will make a negligible change in y, so that we have $f = f_x = 0$. It is not quite so clear why this should also happen at a cusp, but he points out that here $f_x = f_y = 0$.

At a point of this sort the two equations in x,

$$a_0 x^n + a_1 x^{n-1} + \ldots + a_n = 0,$$
$$na_0 x^{n-1} + (n-1)a_1 x^{n-2} + \ldots + a_{n-1} = 0,$$

where a_i is a polynomial in y, will have a common root. The same will be true of

$$a_0 x^{n+(p/q)} + a_1 x^{n+(p/q)-1} + \ldots = 0,$$
$$\left(n+\frac{p}{q}\right)a_0 x^{n+(p/q)-1} + \left(n+\frac{p}{q}-1\right)a_1 x^{n+(p/q)-2} + \ldots = 0.$$

Multiply through by q and divide by $x^{p/q}$:

$$(nq+p)a_0 x^{n-1} + (nq+p-q)a_1 x^{n-2} + \ldots = 0.$$

If therefore we start with our original equation and multiply its terms one by one by the members of any arithmetical progression, and if the new equation and the old one have a common root, the original equation had two equal roots: 'C'est précisément en quoy consiste la méthode de M. Hudde.'[‡] We have seen just this in Hudde's own setting on p. 134.

Let us next find which tangent to a curve passes through the point $(s, 0)$. Let a vertical through the origin meet this at $(0, t)$,

$$\frac{t}{s} = \frac{y}{s-x}, \qquad x = \frac{s(t-y)}{t}.$$

We substitute this value in the equation of the curve, and put down the condition that the resulting equation in y shall have two equal roots, since the tangent has two superposed intersections with the curve. This gives the desired relation between s and t; he remarks that if an equation have a double root this will be a root of the derivative also. He then proceeds to find inflexions, using the same technique, but requiring the equation in y to have three equal roots, so that second differences vanish.

To draw a normal to a curve from the point (s, t) we seek a circle with this point as centre which has two adjacent intersections with the curve. We write

$$(s-x)^2+(t-y)^2 = r^2,$$

$$x = s \pm \sqrt{\{r^2-(t-y)^2\}}.$$

We then substitute and put down the condition that the resulting equation in y shall have two equal roots. This scheme was familiar to Descartes. L'Hospital closes with the remark, p. 181:

'On voit clairement par ce que l'on vient d'expliquer dans cette section de quelle manière l'on doit se servir de la méthode de Mrs Descartes et Hudde pour résoudre ces sortes de questions lorsque les courbes sont géométriques. Mais l'on voit aussi en même temps qu'elle n'est pas comparable à celle de M. Leibnis [sic] que j'ai tâché d'expliquer à fond dans ce traité.'

I have given, I think, a sufficient account of the contents of the *Analyse*; it is time to take up the vexed question of originality. The point at issue is just how much did L'Hospital take direct from John Bernoulli, who spent some time in Paris in 1694, and also at the marquis's country place of Ouques, and beyond a doubt introduced his host into the calculus of Leibniz. This question has been vigorously debated pro and con. Cantor sticks up for the Frenchman, Montucla inclines to the Swiss side. A long discussion will be found in Schafheitlein (q.v.) and Rebel (q.v.). On p. xiv of the Introduction to L'Hospital[1] the writer says:

'Au reste je reconnois devoir beaucoup aux lumières de Mrs Bernoulli, surtout à celles du jeune, présentement professeur à Groningue. Je me suis servi sans façon de leurs découvertes et de celles de M. Leibnis. C'est pourquoy je consens qu'ils revendiquent tout ce qu'il leur plaira, me contentant de ce'qu'ils voudront me laisser.'

This is a sufficiently casual way to treat the question of authorship. L'Hospital sent a copy of the *Analyse* to John Bernoulli in 1697 for which the latter returned thanks and an expression of appreciation. But in 1698 he wrote a letter to Leibniz complaining bitterly that L'Hospital had copied from him shamelessly, and a similar note appeared in a letter to Brooke Taylor written after L'Hospital's death.

The reason for concealing his displeasure in private letters would seem to be found in the fact that Bernoulli was indebted to L'Hospital for his position as professor at Groningen. The Bernoullis were a contentious race, and one does not get a very pleasant impression of John from the incident. My own feeling is that L'Hospital was culpably negligent in acknowledging his scientific indebtedness, but that he did not intentionally take to himself any credit due to another.

Let us actually compare the *Analyse des infiniment petits* with Bernoulli's *Differentialrechnung* from the Basel manuscript translated and annotated by Schafheitlein (q.v.). Bernoulli begins with three postulates. Here are two of them:

'Eine Grösse, die vermindet oder vermehrt wird um eine unendlichkleinere Grosse, wird weder vermindert noch vermehrt.'

'Jede krumme Linie besteht aus unendlich vielen Geraden die selbst unendlich klein sind.'

Here are L'Hospital's corresponding assumptions:

'On demande qu'on puisse prendre indifféremment l'une pour l'autre, deux quantités qui ne diffèrent entre elles que d'une quantité infiniment petite, ou (ce qui est la même chose) qu'une quantité qui n'est augmentée ou diminuée que d'une autre quantité infiniment plus petite qu'elle, puisse être considérée comme demeurant la même.'

'On demande qu'une ligne courbe puisse être considérée comme l'assemblage d'une infinité de lignes droites, chacune infiniment petite, ou (ce qui est la même chose) comme un poligône d'un nombre infini de côtés, chacun infiniment petit, lesquels déterminent par les angles qu'ils font entre eux, la courbure de la ligne.'

The two are essentially identical, L'Hospital's being in each case longer and more explicit.

Bernoulli's first section gives the rules for differentiating sums and differences, products and quotients much as L'Hospital does. At times the actual examples are the same, as $\sqrt{(ax+x^2)}/\sqrt{(x^2+y^2)}$. The second section in each book goes to finding subtangents, each beginning with $ax = y^2$. An interesting problem comes in Bernoulli,† to find a curve whose subtangent is constant:

$$\frac{y\,dx}{dy} = a; \qquad \frac{dx}{a} = \frac{dy}{y}.$$

'Weil aber dx/a stets constant ist, so wird dy/y immer constant, d. h. (die Ordinaten) bilden eine geometrische Reihe und es ist daher die logarithmische Linie, deren Ordinaten eine geometrische und die Abscissen eine arithmetische Reihe bilden.'

† p. 21.

I am at a loss to know why neither writer proved

$$\int \frac{dy}{y} = \log y.$$

Perhaps no good proof was available; none is too easy to find to-day.

Bernoulli's treatment of maxima and minima is less complete than L'Hospital's, but several of the most interesting problems are identical. I mention in particular the day of the longest twilight, and the determination of the lowest point reached by a weight whose cords pass over two pulleys. Not only are the problems the same, but the figures also, except that L'Hospital's are much better drawn. Each writer then passes to points of inflexion, frequently using the same curves. Bernoulli goes no farther, he offers no parallels to L'Hospital's last six sections. In this connexion Schafheitlein makes a devastating remark:

'Aus dem Briefwechsel ergiebt sich dass nun, grössere Teile von Abschnitt 2, 4, und 5, und völlig Abschnitt 8 und 9 von Bernoulli herrühren, die Abschnitte 6, 7 und 10 geben nur Anwendungen der Differentialrechnung auf Dinge, die schon durch Tschirnhaus, Descartes und Hudde grossenteils bekannt waren.'†

I do not know the basis of this statement. The only part of the correspondence I have seen is that given by Rebel, which does not bear the idea out. I do not see as much identity in L'Hospital and Bernoulli as does Schafheitlein. I find it hard to believe that a man so much respected by his contemporaries could have published a book that was almost entirely the work of others.

What is perfectly certain is that L'Hospital published a magnificent text-book, even Bernoulli acknowledged that. If once we accept his postulate about magnitudes of different orders where the lesser can be neglected in comparison with the greater, it is hard to see how a much better presentation could be found. Here is Cantor's opinion of the *Analyse*:

'Der Erfolg des Buches ist um so begreiflicher als es das erste, langer Zeit das einzige, und noch längere Zeit das am leichtesten lesbare Lehrbuch der Differentialrechnung war.'‡

§ 3. *Traité analytique des sections coniques*

I have spoken of L'Hospital as a great text-book writer. His reputation in that matter does not rest alone on the *Analyse* but to an almost equal extent on his analytic study of the conic sections to which I now turn.

The book begins with the parabola, defined mechanically by sliding a right triangle. He works at the equation where the coordinate axes are a tangent and the diameter through the point of contact, proving that

† Schafheitlein (q.v.), p. 6.   ‡ Cantor[1], 1st ed., vol. iii, p. 235.

the abscissa is the negative of the X intercept of the tangent. In Book II we come to the ellipse. He works out the equation in this excellent form. Let the foci be F and f, a point on the curve M, the distance from centre to focus $c = \sqrt{(a^2-b^2)}$:

$$Mf^2 - MF^2 = [y^2+(x+c)^2] - [y^2+(x-c)^2]$$
$$= 4cx$$
$$= (Mf-MF)(Mf+MF)$$
$$= (Mf-MF)2a.$$

$$Mf - MF = \frac{2cx}{a}.$$

$$Mf = a + \frac{cx}{a}, \qquad MF = a - \frac{cx}{a};$$

$$y^2 = MF^2 - (c-x)^2$$
$$= \frac{b^2}{a^2}(a^2-x^2).$$

$$\frac{y^2}{(a+x)(a-x)} = \frac{b^2}{a^2}.$$

A line at the extremity of an axis, parallel to the other axis, will not meet the curve again and so be a tangent. He then proceeds to find conjugate diameters in this curious fashion. Through the centre C draw a line to the point $P = (x_1, y_1)$ on the curve. Find $(x_2, 0)$ so that $x_1 x_2 = a^2$. Then a diameter parallel to the line from (x_1, y_1) to $(x_2, 0)$ will be conjugate to the diameter CP. Of course the double ordinate through (x_1, y_1) lies on the polar of $(x_2, 0)$, as is shown in Apòllonius, iii. 37, and the line from (x_1, y_1) to $(x_2, 0)$ is the tangent at (x_1, y_1). L'Hospital proves that the relation of conjugate diameters is a reciprocal one.

In Book III we reach the hyperbola. I mention in passing that he looks on the two branches as separate curves, calling them 'Hyperboles opposées'. The conjugate axis is found by taking a perpendicular through the centre to the transverse axis, and bringing it to intersect a circle whose centre is an end of the transverse axis, and whose radius is the focal distance from the centre. With a, b, c thus determined we get the asymptotes. The tangent is found to be such a line through a point of a curve that the given point is the middle of the segment cut by the asymptotes. Tangents lead to conjugate diameters.

The fourth book deals with the three conics taken together, though frequently the parabola must be left aside. On p. 95 we have the famous problem of drawing a tangent to a conic from a point outside. In the case of a parabola you draw a diameter through the point. Then the reflection of the given point in the intersection of the diameter and the curve will lie on the polar of the given point, which is parallel to the tangent at the intersection. We have essentially the same construction for the central conics. On p. 102 is a pretty construction for the ellipse,

$$\frac{x^2}{a'^2} + \frac{y^2}{b'^2} = 1.$$

We connect the point $\left(a' + k, \frac{2b'}{a'}\right)$ with the point $(a', 0)$ and the point $(a', -k)$ with the point $(-a', 0)$. The point of intersection of these lines as k varies traces the desired curve. This is a special case of the Chasles-Steiner construction by means of projective pencils; it recalls De Witt's work. He ends by passing a conic through five points. The construction is based on the theorem, known to the Greeks, which he has already proved, that the ratio of the distances from a point to the intersections of the curve with two lines through the point depends on the directions of the lines and not on the situation of the point. The construction is that of Pappus.

In Book V we have a slight smattering of the calculus. We have Cavalieri's principle of equal areas leading to a proof that if we draw parallel chords to an ellipse, and connect each end of one with one end of the other, making an inscribed trapezoid, the areas of the segments outside the non-parallel sides are equal. On p. 156 we come to a more general study of sub-tangents, though the notation is that of differentials.

Suppose we have the curve

$$y^n = x^m a^{n-m}.$$

Let s be the sub-tangent. The ratio of the increments is s/y, so we take these as e, ey/s

$$\left(y + \frac{ey}{s}\right)^n = (x+e)^m a^{n-m},$$

$$ny^{n-1}\frac{ey}{s} = mx^{m-1}ea^{n-m},$$

$$s = \frac{n}{m}x.$$

He then proceeds to find the formula for the area under the curve. The increment of area is ye. The increment of xy is

$$(x+e)\left(y+\frac{e}{s}\right) - xy = ye\left[1+\frac{x}{s}\right].$$

The ratio of the increments is

$$\frac{1}{1+x/s} = \frac{m}{m+n}.$$

Since this is constant, the ratio of the increments is the ratio of the areas.

On p. 164 he develops, very rigorously, a relation between the arc of a parabola, and the arc under a hyperbola. Take the curves

$$y^2 = 2mx; \qquad x'^2 - y^2 = m^2.$$

For the parabola $ds = \dfrac{dy}{m}\sqrt{(m^2+y^2)} = \dfrac{x'\,dy}{m}.$

Hence the length of the arc multiplied by m is equal to the corresponding area under the hyperbola. He uses this much later to find a parabolic arc twice as long as a given one. This example is due to Van Heurac, and will be found in Descartes[2].†

In Book VI the conics are taken up as Apollonius treated them, namely as sections of a circular cone. The words 'ellipse', 'parabola', and 'hyperbola' are used, but it is not until much later that he shows that these curves are the same as those of the same names discussed before. The line where the plane of the circle meets the plane through the vertex parallel to the plane of the conic, the line that is projected to infinity, is called, unfortunately, the directrix. A diameter is defined as the line connecting points of contact of parallel tangents. Much later he shows that all such are concurrent. The whole book seems to me superfluous.

Book VII is entitled *Des lieux géométriques* and deals with the general equation of the second order. There are here two questions:

1) *Given the data necessary to determine a conic, to find its equation.*

2) *Given the equation, to determine the conic.*

With regard to the first question, suppose that we know the major axis of an ellipse in length and position, to find the equation we must have one other datum, say the position of one point (x, y). We then find b from the Greek form of the equation

$$y^2 = \frac{b^2}{a^2}x(2a-x).$$

As for solving the general equation of the second order, that had already been done by Fermat, Descartes, and De Witt; naturally L'Hospital does it more neatly.

Book VIII contains a number of locus problems, of which I will give two.

Problem] *Given a set of rectangular hyperbolas with the same asymptotes to find the locus of the feet of the normals on them from a point (a, b).*

$$xy = c^2; \qquad x\,dy + y\,dx = 0;$$

$$\frac{y-b}{x-a} = \frac{x}{y};$$

$$x^2 - y^2 = ax - by.$$

This is clearly another rectangular hyperbola.

† Vol. i, p. 518.

Problem] *To find the locus of a point which moves so that the tangents thence to a given parabola make a given angle.*

I assume that the angle is not a right angle, in which case the locus would be the directrix. Let the parabola be

$$y^2 = 2mx.$$

The points of contact of the tangents shall be $(x_1 y_1)$, $(x_2 y_2)$, their slopes m/y_1, m/y_2, the tangent of the angle m/n;

$$\frac{\dfrac{m}{y_1} - \dfrac{m}{y_2}}{1 + \dfrac{m^2}{y_1 y_2}} = \frac{m}{n}; \quad n(y_2 - y_1) = y_1 y_2 + m^2.$$

Since (x, y) is on the tangent, say at (ξ, η),

$$y\eta = m(x+\xi) = mx + \frac{\eta^2}{2},$$

$$\eta^2 - 2y\eta + 2mx = 0.$$

The roots in η are y_1 and y_2:

$$y_2 - y_1 = 2\sqrt{(y^2 - 2mx)}$$
$$4n^2(y^2 - 2mx) = m^2(2x+m)^2.$$

L'Hospital's work is a good deal longer than this, for he used neither the formula for the tangent of the difference of two angles, nor those connecting the roots and coefficients of a quadratic equation, though both were known in his time.

Book IX deals with the use of conics to solve higher equations, an exercise dear to Descartes. For instance, on p. 299 we are asked to show how any equation of the fourth order can be solved by the aid of a circle and a parabola

$$x^4 + 2bx^3 + acx^2 - a^2dx - a^3f = 0.$$

This is written in the homogeneous form usual at the time. We begin with the parabola

$$ay = bx + x^2,$$

which is easily constructed. Squaring we have

$$a^2y^2 = x^4 + 2bx^3 + b^2x^2,$$

$$a^2y^2 + (ac - b^2)x^2 - a^2dx - a^3f = 0.$$

We bring them to intersect. The abscissae of the intersections give the desired roots, but through these four points will also pass the circle

$$a^2(x^2+y^2) + (a^2b + b^3 - abc - a^2d)x - a(a^2 + b^2 - ac)y - a^3f = 0.$$

A more ambitious problem comes on p. 336, an equation of the sixth order which, however, lacks the second term. Incidentally I do not know whether L'Hospital knew how to remove this term when it was present

$$x^6 - bx^4 - cx^3 + dx^2 - fx + g = 0.$$

Take

$$a^2y = x^3,$$

$$dx^2 + a^4y^2 - a^2bxy - fx - ca^2y + g = 0.$$

Book X, the last and the longest, deals with determinate problems, which amounts to finding the intersections of pairs of conics. I give some examples.

Problem] *To find a point which has the property that the differences of its distances from three given points have given values.*

This amounts to finding the intersections of two hyperbolas with a common focus, and so, by circular inversion, to finding the common tangents to two given circles. L'Hospital realizes that if we go about this stupidly we shall run into complicated expressions, so he uses ingenuity. The focal distances from a point on the hyperbola

$$\frac{x^2}{a^2} - \frac{y^2}{b^2} = 1$$

are $ex+a$, $ex-a$, where $ae = \sqrt{(a^2+b^2)}$. Here x is the distance from the point midway between A and B to the foot of the perpendicular dropped on AB from the point sought

$$AB = 2ae = 2\sqrt{(a^2+b^2)}; \qquad MA - MB = 2a.$$

Similarly

$$MA = e'x' + a', \qquad MC = e'x' - a',$$

$$MA - MC = 2a', \qquad AC = 2a'e'.$$

We know a, e, a', e' and the position of C. Let AC make an angle θ with AB so that C is the point $(-ae+2a'e'\cos\theta, 2a'e'\sin\theta)$.
The mid-point of AC is $(-ae+a'e'\cos\theta, a'e'\sin\theta)$.

A line perpendicular to AC is $\dfrac{y-\eta}{x-\xi} = -\cot\theta.$

The distance to this line from the mid-point of AC is

$$x' = a'e' - ae\cos\theta - (x\cos\theta + y\sin\theta).$$

Setting

$$ex + a = e'x' + a',$$

we have a straight line whose intersections with the hyperbola gives the points desired. L'Hospital points out the relation to the problem of Apollonius, to construct a circle tangent to three given circles.

Problem] *Given a circle of centre A and two points off the circle E, F, to find such a point M on the circle that* $\angle FMA = \angle AME$.

This is Leonardo's optical problem. L'Hospital gives Huygens's solution which we saw on p. 53.

Problem] *Given a point within a given angle, to pass a circle through the point which shall touch one leg of the angle, and cut a segment of given length on the other leg.*

According to the first requirement the centre of the circle must be on a parabola having the given point as focus and the first leg as directrix. If it pass through a given point and cut a segment of given length on a given line, the difference between the squares of the distances from the centre of the circle to the given point and to the nearest point of the given line will be constant. This will give the equations

$$x^2 + y^2 - (x-b)^2 = a^2,$$
$$y^2 = -2bx + a^2 + b^2,$$

and so another parabola.

The book closes with a long discussion of the problem of dividing an arc of a circle of unit radius into a given number of parts. The basis of the discussion is this formula:

$$2\cos n\theta = 2\cos\theta \times 2\cos(n-1)\theta - 2\cos(n-2)\theta.$$

This he does not prove, but gives examples showing its accuracy up to $n = 13$. An immediate proof comes if we write $2\cos x = e^{ix} + e^{-ix}$. There follow a large number of applications, which seem to include some errors. The book ends with the description of a link work which will divide a given angle into a given number of equal parts.

There remains the question of the originality of the work. I confess that after studying the story of the *Analyse des infiniment petits* it seemed to me quite likely that he had leaned heavily on some previous author. For a moment I suspected plagiarism of Philippe de la Hire's *Nouveaux élémens des sections coniques* which was published in 1679. The two works take up many of the same problems, but in different orders. L'Hospital merges geometric and algebraic methods, La Hire keeps the two separate. I think the charge of undue copying can be dismissed, at least in this case.

L'Hospital's *Traité des sections coniques* is a great text-book. The proofs are uniformly easy; often, one suspects, the easiest possible. As in the previous book he could have saved himself much trouble had he made use of trigonometry, and he might have, with advantage, made further use of the calculus. He probably had the same scruple about doing this which besets modern text-book writers, a feeling that didactically the calculus is a later subject than analytic geometry;

largely a mistake in my opinion. But I repeat, we have here a great text-book. Here is the view of W. W. Rouse Ball:

'He wrote a treatise on analytic geometry which was published in 1707 and for nearly a century was deemed a standard treatise on the subject.'[†]

In the same spirit we find Cantor:

'Das Werk besitzt die gleichen Eigenschaften, welche auch De L'Hospitals Analyse des Infiniment Petits nachzurühmen sind, eine ungemeine Fasslichkeit bei grosser Sorgfalt zahlreiche Einzelsätze zu geben. Bahnbrechende Neuerungen sind freilich nicht gar zu erwarten.'[‡]

[†] Ball (q.v.), p. 380.   [‡] Cantor, 1st ed., vol. iii, p. 410.

CHAPTER XIII

BUFFON

§ 1. The search for the infinite

THERE certainly never was a man belonging to that class which I have called amateur mathematicians who had a wider interest in all science, especially descriptive science, than George-Louis Leclerc, Comte de Buffon. His monumental *Histoire naturelle* is overwhelming in size and the variety of topics treated, but he had besides a very real interest in theoretical mathematics. He was especially preoccupied with what one might call the metaphysical aspect of the subject. The discovery of the infinitesimal calculus had raised a number of very puzzling questions as to the real meaning of the various concepts involved. What do we really mean by infinitely large or infinitely small ? What, if anything, is indivisible ? The fundamental concept of a limit was only reached after long and painful struggle. These abstruse questions appealed strongly to Buffon's inquiring temper although, like his contemporaries, he never reached a point where he could give answers which are entirely satisfactory to our more critical minds.

Buffon's first mathematical effort was a translation of Newton's *Fluxions*. He did not base this on the Latin original, but on the English translation by Colson. However, he wisely confined himself to giving nothing but Newton, omitting Colson's long commentaries.

Buffon is very enthusiastic about the genius of the original:

'On sera bien aise de voir, en un seul petit volume, le calcul différentiel et le calcul intégral avec toutes leurs implications. On reconnaîtra, à la manière dont les sujets sont traités, la main du grand maître, et le génie de l'inventeur, et on demeurera convaincu que Newton seul est l'auteur de ces merveilleux calculs, comme il est aussi de bien d'autres productions toutes aussi merveilleuses.'†

After this he proceeds to slay Leibniz, and such of his disciples as claim that he was the author, or co-author, of the calculus. He makes out a pretty good case too, although he hardly lets us see that whatever else Leibniz was or was not, he was an extremely great mathematician.

Buffon takes up the quest of the infinite, about which he had very definite ideas, which I must confess are not perfectly clear to me. He feels that the Greeks understood perfectly well what infinity means but that more recent writers have invented perfectly new concepts which are wrong. To Buffon something is infinite if it lacks a bound or a last term. Then we find the curious statement:

'Ce n'est pas ici le lieu de faire voir que l'espace, le temps, la durée ne sont

† Buffon[1], p. vi.

pas des infinis réels, il nous suffira de prouver qu'il n'y a point de nombre actuellement infini ou infiniment petit. . . . Mais dira-t-on le dernier terme de la suite 1, 2, 3, 4,..., n'est il pas infini ? . . . Il paroit que les nombres doivent à la fin devenir infinis puisqu'ils sont toujours susceptibles d'augmentation ; à cela je réponds que cette augmentation, dont ils sont susceptibles, prouve évidemment qu'ils ne peuvent être infinis . . . mais que tous sont de même nature que les précédents, c'est à dire, tous finis, tous composés d'unités.'†

I confess that I do not know just what he is driving at here. There seems to be what we should call a confusion between cardinal and ordinal infinite. His adversary objects that the numbers are cardinally infinite, but he replies by saying that there is no ordinal infinite, as whenever we add a number it is the same sort of thing as what precedes ; we never get out of the system by adding, hence it is self-contained or bounded. These ideas were very dear to Buffon, I do not think they are very important to us to-day.

§ 2. Moral arithmetic

Buffon's principal mathematical work was his *Essai d'arithmétique morale*. This begins with a discussion of certainty. There are two kinds of certainty. Physical certainty is based on a long uninterrupted series of successes ; moral certainty, which is much less strong, is based on a restricted number of observations, of what seem essentially analogous cases. And then he undertakes the curious task of calculating the probability that the sun will rise the next day. This is obviously a very foolish question. Bertrand has pointed out that a man might be con-vinced that the sun was sure to rise the next morning, and then travel to the polar regions where the phenomenon failed. Even if we made a mathematical calculation, the correct way would be to go on Bayes's principle of the probability of causes. Buffon assumes that after each day when the sun has risen the probability that it will rise the next day is doubled. Assuming that it has already risen 2,190,000 times, the probability that it will rise the next day is

$$1 - \frac{1}{2^{2,190,000}}.$$

When it comes to moral certainty, he takes as his basis the assumption that a man is morally certain he will be alive the next day, and so a thing is morally certain if the chance of failure is less than 1/10,000.

The title of the essay is *Arithmétique morale*, and he pays a good deal of attention to the moral value of money, which is not the same as its arithmetical value. The arithmetical value of ten crowns is the same to all men, but the importance of an accession of ten crowns is very

† Buffon[1], pp. ix, x.

different to a man who has only five and to one who has five million. He argues ingeniously from this to show the foolishness of games of chance. Suppose we have two players, each with a certain sum. They agree to play at a fair game until one has lost one-half of that sum. At the end the winner will have gained one-third of what he then has but the loser will have lost one-half of his original property. Buffon argues that there is an adverse chance of one-sixth. I can see no justification for estimating the value of the sum in one case in comparison with what he had at the end and in the other case on what he had in the beginning ; it would be more logical to say that the winner had gained what was one-third of what he then had while the loser had lost 100 per cent. of what he still owned. But I approve of his disapproval of gambling.

Buffon passes next to the famous paradox of St. Petersburg :

'Cette question a été proposée pour la première fois par feu M. Cramer, célèbre professeur de mathématiques à Genève dans un voyage que je fis à cette ville en l'année 1730. Il me dit qu'il avoit été proposé précédemment par M. Nicolas Bernoulli à M. de Montmort . . . je rêvai quelque temps à cette question sans en trouver le nœud, je ne voyois pas qu'il fût possible d'accorder le calcul mathématique avec le bon sens, sans y faire entrer quelques con- sidérations morales, et ayant fait part de mes idées à M. Cramer il me dit que j'avais raison, et qu'il avoit aussi résolu cette question par une voie semblable ; il me montra ensuite sa solution à peu pres telle qu'on l'a imprimée depuis dans les Mémoires de l'Académie de Pétersbourg 1738.'†

Here is the paradox. Peter and Paul play under these rules. Paul throws a coin. If it come up heads he will give Peter a crown. If it come up tails he will not give anything, but will throw again. If it come up heads the second time he will give two crowns, if the first heads be on the third throw he will give 4 crowns, if not till the nth throw he will give 2^{n-1} crowns. What should Peter pay to be allowed to take part in this interesting game ?

We have here a question of mathematical expectation, that being by definition the sum of all possible gains each multiplied by the chance of getting it. Peter's expectation is, then,

$$\frac{1}{2} \times 1 + \frac{1}{2^2} \times 2 + \frac{1}{2^3} \times 2^2 + \dots = \infty.$$

Now this is certainly contrary to common sense. No one with any common sense would pay any large sum for Peter's expectation. What is the matter ? Buffon had the bright idea of having a child throw a coin 2,048 times. The total payment was 10,057 crowns, so that the average was about 5. Peter should offer about 5 crowns to come into the game, not an infinite sum.

† Buffon², p. 75.

Buffon argues rightly that the game is inherently impossible. Neither adversary will have an infinite fortune, so that Paul is in no position to meet all possible outcomes. Moreover, there is an obvious limit of time ; no matter how industriously they play they will not be able to keep on long enough to permit more than moderate gains on Peter's part. He then gives his own explanation. The moral value of an accession of money depends, not only on the amount, but also on how much one has at the start. He suggests that the sum given if the first head come on the nth throw should not be 2^{n-1} but $(9/5)^{n-1}$. Peter's expectation should then be

$$\frac{1}{2} \times 1 + \frac{1}{2^2} \times \left(\frac{9}{5}\right) + \frac{1}{2^3} \times \left(\frac{9}{5}\right)^2 + \ldots = 5.$$

The assumption seems to me perfectly arbitrary, one cannot escape the impression that it was made to work out to the answer 5.

This Petersburg paradox has been quite famous in the history of mathematics. The idea of the moral value of money was developed by Daniel Bernoulli. He maintained that if a man have a fortune of x, the moral value of a small increment of dx, is dx/x. Hence the moral value of an increase of one's fortune from a to b is

$$\int_a^b \frac{dx}{x} = \log\frac{b}{a}.$$

A system of direct taxation where each surrenders the same proportion of his wealth will thus impose the same moral strain on each.

§ 3. Geometrical probability

Buffon's most original contribution to mathematics was the invention of geometrical probability, which seems to have been entirely the child of his brain.

'L'analyse est le seul instrument dont on se soit servi jusqu'à ce jour, dans la science des probabilités, pour déterminer et fixer les rapports du hasard, la géometrie paraissoit peu propre à un ouvrage aussi délié ; cependant si l'on y regarde de près, il est facile de reconnoître que cet avantage de l'analyse à la géométrie est tout à fait accidentel.'†

He proceeds to invent some examples of geometrical probability. A smoothed table is ruled with squares. A coin, whose diameter is less than the width of these squares, is thrown at random on the table ; what is the probability that it will cross the boundary of one of the squares ?

He assumes that all situations for the centre of the coin are equally likely. The probability is the proportion to the area of the square of the area of that part in which the centre must lie in order to cross.

† Buffon², p. 96.

If the length of a side of a square is 1, and if the answer is to come out $\frac{1}{2}$, then in order to cross, the centre must lie inside a strip running around inside the boundary with a width equal to the radius of the coin. The strip must have an area equal to one-half of the area of the square; hence the side of the square must stand to the diameter of the coin in the ratio

$$1 : (1 - \sqrt{\tfrac{1}{2}}).$$

He discusses certain modifications of this problem when the table is ruled with equal figures of different shapes. Then comes the problem which has made Buffon's name famous in the history of probability.

Problem] *A floor is ruled with parallel lines whose distance apart is d. A needle of length $l < d$ is thrown at random on the floor. What is the probability that it will cross one of the lines?*†

In order to cross a line the needle must fulfil two conditions. The distance from the centre to the nearest line must be less than half the length of the needle, and the angle which the needle makes with a perpendicular to the lines must be less than $\cos^{-1}(2x/l)$. These are treated as independent probabilities, so the final probability is their product. Assuming that all distances to the nearest line are equally likely and that all angles are equally likely, we have

$$\frac{4}{\pi d} \int_0^{\frac{1}{2}l} \cos^{-1} \frac{2x}{l} \, dx = \frac{2l}{\pi d}.$$

There are various things which should be said about this. Most important is that Buffon did not perceive the great dangers involved in assuming the equally likely, or the choice of independent variable. His assumptions are the most plausible ones, but the case is not always so simple. Suppose, for instance, we ask this question: a number is taken at random between 1 and 3, what is the probability that it lies between 1 and 2. The distance between 1 and 2 is one-half the distance between 1 and 3. Hence the most natural answer is that the probability is $\frac{1}{2}$. But wait a moment. If a number is between 1 and 3, its reciprocal is between 1 and $\frac{1}{3}$, while if it is between 1 and 2, its reciprocal is between 1 and $\frac{1}{2}$. If thus we fix our attention on the reciprocal, the question is thus: A number is between 1 and $\frac{1}{3}$, what is the probability that it lies between 1 and $\frac{1}{2}$? The natural answer here is $\frac{3}{4}$.

I should mention next that it does not seem that Buffon was quite up to integrating $\cos^{-1}(2x/l)$. He notes that the integrand is like what appears when we are seeking the area under a cycloidal arch, and that he knows, so that he gives his answer in those terms. It is also interesting to note that we can get the answer without any integration by a

† Ibid., p. 101.

device described on p. 56 of Bertrand's *Calcul des probabilités*, and ascribed to Rabier.

The probability is the expectation of a man who is to receive one crown if the crossing takes place. This expectation is the sum of the expectations of the various linear elements of the needle, and these are unaltered if the needle is bent to a circular shape. The probability of crossing is now the ratio of the diameter of this circle to $d/2$, but if the bent needle cross a line once, it will cross it twice, so that the expectation is twice the new probability. This will give the same answer as before.

The needle problem has been studied experimentally by a number of persons, partly as an amusing method of calculating the value of π. The most apparently successful attempt I have heard of was made by Lazzerini in 1901, with the result

$$\pi = 3\cdot1415929...,$$

an error of $0\cdot0000003$. This is certainly suspiciously accurate. It would only be possible if the number of crossings were exactly the most likely with no discrepancy, but the chance for that to happen is approximately 1/69, so that it seems quite likely that Lazzerini 'watched his step' and stopped his experiment at the moment when he got a good result.[†]

Buffon also considers the problem when the floor is covered with rectangles of dimensions a and b. Here we have the probability of crossing parallels a apart, plus the probability of crossing parallels b apart, less the probability of crossing a corner. This is a little more complicated than the other, but leads to the same integrals, and the final answer is

$$\frac{2l(a+b)-l^2}{\pi ab}.$$

I do not understand Buffon's symbolism here, and so cannot be sure that he has the correct answer.

In the last section of the *Arithmétique morale* Buffon discusses arithmetical and geometrical measures. He compares our decimal system of notation with other scales. Suppose we wish to express in the quinary scale

$$1738 = a\times5^n; \qquad \frac{\log 1738}{\log 5} = 4+.... .$$

$$1738 = 2\times5^4+488$$
$$488 = 3\times5^3+113$$
$$113 = 4\times5^2+13$$
$$13 = 2\times5+3.$$

The answer is 23423.

[†] Cf. Coolidge[2], pp. 81, 82.

When it comes to geometrical measures he is on more dangerous ground. We can measure line segments, that is to say their lengths, by means of a unit of length, but measuring curved lengths involves the use of infinitely small lengths, whatever they may be. He sticks to the old distinction between geometrical and mechanical curves. He realizes that infinitely small quantities are only useful when we compare them, that is to say take their quotients or add an infinite number. The concept of an infinitesimal as a variable, approaching 0 as a limit, not a fixed quantity, was still far in the future. He insists on the fact that it is indeed lamentable that there are such a variety of measures differing from country to country, and points out the advantages of having a uniform system for all, based on some physical fact, independent of country. He suggests as a unit of length that of a seconds pendulum at the equator. This unit had been suggested by Picard in the seventeenth century, and was considered by the committee which devised the metric system, but was discarded in favour of the metre which was supposed to be one ten-millionth of the meridian of Paris. Curiously enough this is close to the length of the seconds pendulum, but why go to the equator for a unit when you can have one in Paris ? As a matter of fact the official metre is not exactly equal to either.

We cannot look on Buffon as a great mathematician, but a bright man, interested in mathematics, who deserved well of posterity.

CHAPTER XIV

DENIS DIDEROT

§ 1. Vibrating strings

FEW persons think of this intellectually omnivorous encyclopaedist as a mathematician, and he certainly was not a mathematician of the first rank. But he was a man of a keen and inquiring mind, who in his earlier years was really interested in mathematics, especially in their applications; yet his discussion of the current theories of mathematical probability, which I personally have not seen, show an equal interest in the philosophy of the subject. I will discuss the mathematical parts of Diderot (q.v.) which was reviewed at some length in the April 1749 *Mémoires de Trévoux pour l'Histoire des Sciences et des Beaux Arts*. The reviewer was principally interested in Diderot's ideas about music and acoustics, and showed no ability to discuss his mathematics. A much better discussion of this is found in Krakauer and Krueger (q.v.).

The first, as well as the longest and most important, memoir in Diderot (q.v.) is entitled 'Principes généraux d'acoustique'. The author's object is to analyse the principles of sound, mathematically and physically, and trace their connexion with music, and the pleasures of listening to it. But how is sound produced? By vibrations of the air, and these in turn are caused by vibrations of a tube or a stretched string; so he comes first to a vibrating string. Here he follows an Englishman named Taylor, of whom I know nothing except that Diderot says he was a contemporary of Newton. This must surely have been Brooke Taylor. We find this theorem:

Lemma II] *The acceleration of any point P of a stretched elastic string of uniform diameter is, during small vibrations, proportional to the curvature of the string at this point.*

We may prove this as follows:

Let the maximum extension of the middle point of the string be $BD = a$, the radius of curvature at this point r. Let

$$MB = x, \qquad SP = a-x,$$

the amount of extension of the general point, whose distance from the centre line is $PM = y$. I apologize to the reader for using y for a horizontal and x for a vertical line, but I thought it best to follow Diderot's notation. The tension at this point shall be G. The length of the chord is L, the mass M. Then by the fundamental principles of mechanics

$$\frac{M \, ds}{L} \frac{d^2 y}{dt^2} = g[(G+\Delta G)\cos(\theta+\Delta\theta) - G\cos\theta],$$

$$-\frac{M \, ds}{L} \frac{d^2 x}{dt^2} = g\left[(G+\Delta G)\sin(\theta+\Delta\theta) - G\sin\theta - \frac{M \, ds}{L}\right]. \tag{1}$$

Taylor assumes that the tension is constant, so that $\Delta G = 0$. He also discards the last term which is the contribution of the mass of an element of length. Expanding we have

$$\frac{M}{L}\frac{d^2y}{dt^2} = -gG\sin\theta\,\frac{d\theta}{ds}; \qquad -\frac{M}{L}\frac{d^2x}{dt^2} = gG\cos\theta\,\frac{d\theta}{ds}. \qquad (2)$$

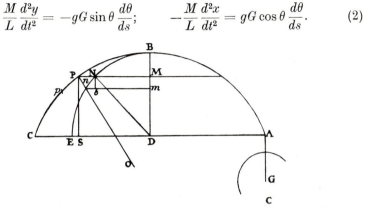

Fɪɢ. 55

We thus get for the normal acceleration

$$-\frac{gGL}{M}\frac{d\theta}{ds},$$

which is proportional to the curvature. This is Taylor's Lemma II.

Proposition 1] *If a string be stretched along such a curve that the curvature at each point is proportional to its distance from the chord, all points will come to lie on the chord at once.*

Here he tacitly assumes that the acceleration is normal to the chord:

$$\frac{d^2(a-x)}{dt^2} = \frac{gaGL}{rM} = k(a-x). \qquad (3)$$

This is the familiar equation for a longitudinally stretched string which comes to rest at the same time, no matter how far it has been stretched. If the string be once stretched along such a curve it will always stay so. It is now time to find the equation of such a curve.†

$$\frac{d^2y}{ds\,dx} = \frac{1}{ra}(x-a), \qquad (4)$$

$$ra\frac{dy}{ds} = \tfrac{1}{2}xx - ax + G.$$

When $x = 0$, $dy = ds$,

$$ra\frac{dy}{ds} = \tfrac{1}{2}xx - ax + ra. \qquad (5)$$

† Diderot (q.v.), p. 24.

Let
$$ax - \tfrac{1}{2}xx = zz;$$
$$dx^2 + dy^2 = ds^2,$$
$$(2razz - z^4)\,dy^2 = (ra - zz)^2\,dx^2.$$

But zz is small compared to ra, so he contracts to

$$\frac{dy}{dx} = \frac{\sqrt{(ra)}}{\sqrt{(2ax - xx)}} \tag{6}$$

$$y = a\sqrt{\frac{r}{a}}\,\cos^{-1}\!\left(\frac{a-x}{a}\right). \tag{7}$$

Diderot notes in this connexion that the differential equation for the motion of a point in the string is essentially that of the weight of a cycloidal pendulum, or a point descending a cycloidal path under gravity where the effective force is proportional to the distance along the curve to the bottom, and from this he calculates the relation between the tension in the string to the data of the cycloid.

These purely mathematical calculations are followed by a long musical discussion which I am unable to criticize. The next piece of pure mathematics comes on p. 62, where he seeks the greatest velocity of any point on the string, this being the velocity of the middle point when it reaches the chord. The integral of equation (3) is clearly

$$\frac{d(a-x)}{dt} = \sqrt{\left(\frac{gaG}{rM}\right)x}. \tag{8}$$

Returning to (7):

when $\qquad y = \dfrac{L}{2}, \quad \cos^{-1}\dfrac{a-x}{a} = \dfrac{\pi}{2}; \quad a\sqrt{\dfrac{r}{a}} = \dfrac{L}{\pi}.$

The maximum velocity comes when $x = 0$. Diderot suppresses the factor g presumably by taking the proper choice of units. We have

$$u = \frac{\pi a\sqrt{G}}{\sqrt{(ML)}}. \tag{9}$$

Problem 1] *A string is stretched by a total force F which gives to the centre the velocity u and to every point such a velocity that at every point the string will always preserve the required curve. What is the value of the maximum displacement?*

The initial velocities must be proportional to the distances $SP = a - x$ if all are to reach the final form together. The initial velocity at D shall be u. He takes the mass of an element as $(M/L)\,dy$. The product of mass and initial velocity is

$$\frac{u(a-x)}{a} \cdot \frac{M\,dy}{L}.$$

Substituting in (6) we get

$$\frac{u(a-x)}{a} \cdot \frac{M\sqrt{(ra)}}{L\sqrt{(2ax-x^2)}}.$$

If a constant force act through a short period the velocity it will produce will be proportional to the force; hence if we integrate with regard to x, whose limits are a and 0, we have

$$\frac{u\sqrt{(ra)}M}{L}.$$

Double this will be the total force F.

$$u = \frac{\pi a \sqrt{G}}{\sqrt{(ML)}}; \qquad \frac{LF}{2M} = \frac{\pi a \sqrt{G}\sqrt{(ra)}}{\sqrt{(ML)}}; \qquad a = \frac{F\sqrt{L}}{2\sqrt{(MG)}}. \qquad (10)$$

The remainder of the first memoir has little mathematics, except for the study of fretting the string in places. The interest is more musical than mathematical.

§ 2. Involutes

Diderot's second memoir is purely geometrical. He is an ardent advocate of Descartes's thesis that it is a great mistake in geometry to

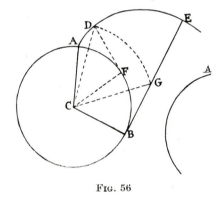

FIG. 56

limit ourselves to what can be accomplished with the sole aid of ruler and compass; on the contrary, we should make use of any curve that can be drawn by mechanical means. The curve he favours especially is the involute of a circle which is easily constructed by unwinding a thread which has been wound around a flat circular cylinder. In fact on p. 121 of Diderot (q.v.) is a picture of a man holding a large circular disk against a wall with one hand, and unwinding the thread with the other. Let the radius of the disk be r, the centre C, the point where the involute springs from the circle A, B any point on the circle, E the corresponding point of the involute. Let A have the polar coordinates

$(r, 0)$. Let positive measure be counter-clockwise, let $\angle ACB = \phi$. The coordinates of E will be

$$x = r \sin \phi - r\phi \cos \phi, \qquad y = r \cos \phi + r\phi \sin \phi. \qquad (11)$$

Diderot does not actually give these equations, but they will somewhat simplify part of what follows. I have copied only one of his figures, there are several on the same page, and he has a confusing trick of changing notation from one to another. Here are what seem to me his most interesting problems, the numbering is mine.

Problem 2] *To divide a given angle in a given ratio.*

Let ϕ be the given angle, the given ratio m/n. Let G divide BE in that ratio. Let a circle of centre C and radius CG meet the involute in D; draw the tangent DF:

$$\frac{BG}{GE} = \frac{m}{n}; \qquad \frac{\widehat{AF}}{\widehat{FB}} = \frac{m}{n}.$$

Problem 3] *To find a sector of a given circle whose area is equal to c^2.*

We wish
$$\tfrac{1}{2}r^2\phi = c^2; \qquad r\phi = \frac{2c^2}{r}.$$

On any tangent to the circle draw a length $2c^2/r$. Find a point on the involute at the same distance from the centre as the end of this tangent. A tangent thence to the circle will determine the required sector.

Problem 4] *To find a square equivalent to a given sector.*

This, by (11), consists in finding a square equivalent to $\triangle CBE$.

Problem 5] *To find the length of an arc AE of the involute.*

For the involute we have $ds = r\phi \, d\phi$, $s = \tfrac{1}{2}r\phi^2$.

Through E draw a perpendicular to EC and let this meet CB in K:

$$BE^2 = CB \cdot BK, \qquad r^2\phi^2 = r \times BK,$$
$$s = \tfrac{1}{2}BK.$$

Problem 6] *To find the area of the involute segment ABE.*

Let us imagine the area filled by infinitesimal triangles bounded by adjacent tangents

$$df = \tfrac{1}{2}r^2\phi^2 \, d\phi; \qquad f = \tfrac{1}{6}r^2\phi^3.$$

The area of $\triangle EBK = \tfrac{1}{2}EB \times BK = \tfrac{1}{2}r^2\phi^3$.

Hence one-third of this area gives the desired result.

Problem 7] *To find the centre of gravity of a circular arc.*

If the central angle be 2θ, we find by a simple integration that the distance along the bisector of the angle to the centre of gravity is

$$d = \frac{r \sin \theta}{\theta} = \frac{r^2 \sin \theta}{r\theta}; \qquad \frac{2r\theta}{r} = \frac{2r \sin \theta}{d}.$$

The distance is, thus, a fourth proportional to the distance along the tangent to the involute, the radius, and the chord.

Diderot next discusses the solution of cubic equations such as

$$x^3 - px = \pm q,$$

which can be at once reduced to the problem of trisecting an angle.

The memoir ends with a long article leading to this:† 'Donc le rayon de la développée est toujours comme l'arc infiniment petit multiplié par le rapport du sinus de l'angle de contingence au sinus versus du même angle.'

The phrase *est toujours comme* here is a little ambiguous, there is a factor 2 which comes in here; the formula he is looking for is

$$r = \frac{2\,ds}{2\,d\theta} = 2\,ds\ \text{ctn}\ 2d\theta = \frac{2\,ds \sin d\theta}{1 - \cos d\theta}.$$

§ 3. The retarded pendulum

Diderot's last memoir of a mathematical nature is No. 5. He investigates the motion of a pendulum in air, assuming that the resistance varies as the square of the velocity. Let the length of the pendulum be a, the height of the top of the swing above the lowest point b, the variable height x, the weight divided by the acceleration of gravity p, the velocity v, the resistance $(fv^2)/q^2$.

The acceleration along the path is

$$\frac{p\sqrt{(2ax - x^2)}}{a} - \frac{fv^2}{q^2} = \frac{dv}{dt} = v\frac{dv}{ds}; \qquad ds = \frac{-a\,dx}{\sqrt{(2ax - x^2)}};$$

$$-p\,dx + \frac{afv^2\,dx}{q^2\sqrt{(2ax - x^2)}} = v\,dv;$$

$$v^2 = 2p(b - x) + 2\int \frac{afv^2\,dx}{q^2\sqrt{(2ax - x^2)}}.$$

Diderot approximates to the remaining integral as follows. If the pendulum were swinging *in vacuo* we should have

$$v^2 = 2p(b - x),$$

but the resistance of the air is small in comparison when the pendulum bob is heavy, hence he rejects this first part except to substitute in the second. He has some difficulty with

$$\frac{4paf}{q^2} \int \frac{(b - x)\,dx}{\sqrt{(2ax - x^2)}},$$

the indefinite integral is rather a complicated expression involving x

† p. 161.

and $\sin^{-1}\sqrt{(2ax-x^2)}/a$. This latter will be proportional to the arc through which the pendulum must swing to reach the lowest point.

The fact is that Diderot had hold of a problem that was too much for him. Krakauer and Krueger go wrong at this point. They write:

'The latter [Newton] had "proved" that the retardation of a pendulum due to the resistance of air in falling through an arc is proportional to the arc. Diderot "proves" it is proportional to the square of the arc' [and in a footnote at the bottom of the same page] 'According to modern physics neither is correct; the resistance varies as the square of the velocity. However, both Diderot and Newton used approximations and arrived at a fairly close result considering the velocity studied, and the short arc traversed.'†

But we have seen that Diderot started by assuming the resistance proportional to the square of the velocity. As for the statement about modern physics, I refer to the long article by Furtwängler.‡ Here it is shown that, physically considered, the resistance of the air cannot be adequately handled in any such summary way. If we put

$$\phi = \sin^{-1}\frac{\sqrt{(2ax-x^2)}}{a}.$$

The classical equation for the pendulum *in vacuo* is

$$\frac{d^2\phi}{dt^2}+\frac{g}{l}\sin\phi = 0.$$

When the amplitude of the swing is small we may replace $\sin\phi$ by ϕ, getting

$$\frac{d^2\phi}{dt^2}+\frac{g}{l}\phi = 0.$$

If we take in the resistance of the air a good approximation is

$$\frac{d^2\phi}{dt^2}+k\frac{d\phi}{dt}+\frac{g}{l}\phi = 0.$$

The retardation of the air is thus proportional to $d\phi/dt$, that is to say the angular velocity. There is a long discussion in Furtwängler's article, or in any good book on rigid dynamics, e.g. Routh, p. 77. Diderot's discussion of Newton, which forms the close of the article, puts the matter in a simpler form than does the master, but both are far from the modern treatment, so it is not worth while following them farther.

Diderot wrote also on 'probability'. This I have not seen. Krakauer and Krueger say, rather obscurely, 'It is regrettable that Diderot did not publish this article. I have the impression it came out after his

† Krakauer and Krueger (q.v.), p. 227.
‡ *Encyclopädie der math. Wissenschaften*, Part IV, Section 7, pp. 14–16.

death.'† It is also possible that he contributed mathematical articles to his *Grande Encyclopédie*, but as the official announcement says at the beginning that the mathematical part was due to D'Alembert it seems needless to try to pick out what might have been contributed by a much inferior mathematician. I cannot leave Diderot without expressing my admiration for his really stimulating mathematical work, when his other interests were so large and so varied.

† Krakauer and Krueger (q.v.), p. 225.

CHAPTER XV

WILLIAM GEORGE HORNER

§ 1. Horner's method

I should perhaps begin with an apology for including this man in my list of distinguished amateurs who have contributed to mathematical science. A great mathematician he certainly was not. A man distinguished in public life he was not, a master in an altogether undistinguished boys' school. The contribution to mathematics for which he is generally known, Horner's method for the solution of numerical equations, was, as we shall see, discovered sixteen years earlier by Paolo Ruffini, and six centuries earlier by Ch'in Chiu-Shao. So why include him ? Well, it may be said, on the other hand, that he was largely self-taught. He never went to the university, but at the age of nineteen became a master in the Kingswood School of Bristol, where he had studied. He read the works of Euler, Lagrange, Halley, and others; he certainly never heard of the work of either of these predecessors. He offers a fine example of what an amateur can accomplish by dogged industry, and his method is surely the best we have for solving numerical equations.

Horner's first article, Horner[1], is a trivial affair, composed, I should judge, about the same time as Horner[2]. This latter has a curious history.

'Mr. Horner's first paper on equations was printed in the Philosophical Transactions in 1819, and Mr. Horner has often stated to me that much demur was made to the inclusion of it in that publication. It was, indeed, owing more to the influence and earnestness of Mr. Davis Gilbert than to any respect for the author, his subject, or his mode of treating it, that the honour was accorded him. The elementary character of the subject was the professed objection, his recondite mode of treatment, was the professed passport of admission.

'The paper has been reprinted in the *Ladies' Diary* for 1838 and most of our readers are doubtless acquainted with it. The mode in which it is drawn up is, in one respect, fortunate, there can be no doubt, since that finally secured its publication, whilst, on the other hand, it may be considered unfortunate by its requiring so much higher mathematical learning to understand the reasoning than the nature of the inquiry itself renders desirable. Mr. Horner was himself so sensible of this objection, that he attempted a simplification of the principles. The consequence of this attempt was the paper now about to be submitted to the public for the first time, after lying more than twenty years altogether unknown.'†

With regard to dates, Horner[2] was presented to the Royal Society in 1819, the year of publication of Horner[1]; Horner[3] was presented to the

† T. S. Davies in a note to Horner[3], p. 108.

Royal Society in 1823, but was inserted in *The Mathematician* for 1845, and finally published in 1855. As for the 'higher mathematical learning' that was so discouraging, that was Taylor's theorem.

Let us now look at Horner[2]:

'The process which it is the object of this essay to establish being nothing else than the leading theorem of the Calculus of Derivation presented under a new aspect, which may be regarded as a universal instrument of calculation, extending to the composition as well as the analysis of functions of every kind. . . .

'In the general equation

$$\phi x = 0$$

I assume

$$x = R + r + r' + r'' + \dots$$

and preserve the binomial character of the operations by making successively

$$\begin{aligned} x = R + z &= R + r + z' \\ &= R' + z' = R' + r' + r'' \\ &= R'' + z'' = \dots \end{aligned} \tag{1}$$

.   .   .   .

By Taylor's theorem, expressed in the more convenient manner of Abrogast, we have

$$\begin{aligned} \phi x &= \phi(R + z) \\ &= \phi R + D\phi R \cdot z + D^2 \phi R \cdot z^2 + \dots \end{aligned} \tag{2}$$

where by $D^n \phi R$ is understood†

$$\frac{d^n \phi R}{1 \cdot 2 \cdot 3 \dots n \, dR^n},$$

He then sets about calculating his derivatives

$$\begin{aligned} \phi R' &= \phi R + Ar \\ A &= D\phi R + Br \\ B &= D^2 \phi R + Cr \end{aligned} \tag{3}$$

.   .   .   .   .   .

$$\begin{aligned} V &= D^{n-2} \phi R + Ur \\ U &= D^{n-1} \phi R + r. \end{aligned}$$

He then writes in reverse:

$$V = D^{n-2} \phi R + D^{n-1} \phi R \cdot r + r^2$$

.   .   .   .   .   .

Keeping on in this way he eventually gets back to (2) with z changed to r.

I note at this point that Horner insists that his method is applicable to all sorts of functions, so that it is not quite clear why r should have

† Horner[2], pp. 49, 50.

the coefficient 1 in the last equation. Having found $\phi R'$ he proceeds to find

$$\phi R'' = \phi R' + A'r'$$
$$A' = D\phi R' + B'r'$$

. . . . . .

If we keep on in this fashion it is well to have a rule for compounding the operator D. Horner writes:

$$D^r D^s \phi a = \frac{(r+1)(r+2)...(r+s)}{1...2...s} D^{r+s}\phi a, \qquad (4)$$

$$D^n \phi R' = D^n \phi R + \frac{n+1}{1} D^{n+1}\phi R.r + \frac{(n+1)(n+2)}{1...2} D^{n+2}\phi R.r^2.... \qquad (5)$$

From this he finally gets an elaborate rule which I will not write, as it does not seem to be of much practical use. He uses the scarcely English word 'derivee' whose nature is well known: 'It is sufficient to state that they may be considered either as differential coefficients, or as the limiting equations of Newton, or as numerical coefficients in the transformed equation.'†

I note that he also uses the word 'derivate'. On the same page he mentions what he calls De Gua's rule that if the roots of ϕ are all real $D^{m-1}\phi$ and $D^{m+1}\phi$ will have opposite signs for each root of $D^m\phi$.

Let us now take some examples of his method, of which we find no simple explanation till we get to Horner[3], which I shall take up later. The fundamental idea is perfectly simple. If we make a first approximation to a root and reduce all of the roots of the equation by this amount, which is easily done by replacing x by $x+r$, and calculate the new coefficients by his method, we have a new equation with roots close to 0, and we can approach these as nearly as we please by repeating the process. When he wishes to avoid decimals he multiplies all roots by 10. His first equation is from Euler:

$$x^4 - 4x^3 + 8x^2 - 16x + 20 = 0.$$

Let us reduce the roots by unity, and then do the same thing again, writing the coefficients downwards in reverse order. We get:

$x =$	20	9	4
	−16	−8	0
	8	2	8
	−4	0	4
	1	1	1.

The first row gives the values of f, the second of Df.

† Horner[2], p. 57.

In the first column are no permanencies of sign, hence the equation has no negative roots. In the third column if we treat 0 as positive the reduced equation has no positive roots; hence the original one had no positive roots greater than 2. He concludes that as the once reduced equation is close to $2x^2+1 = 0$ it has no real roots.

The next equation is taken from Lagrange:

$$x^3-7x+7 = 0$$

$$\begin{array}{ccc} 7 & 1 & 1 \\ -7 & -4 & 5 \\ 0 & 3 & 6 \\ 1 & 1 & 1. \end{array}$$

We see from the last column that there are no positive roots greater than 2. We see by the second column that the first derivative vanishes between 1 and 2; there may be two roots of f in that vicinity. He makes a third table:

$x =$	1·0	1·1	1·2	1·3	1·4	1·5	1·6	1·7
$f =$	1000	631	328	97	−56	−125	−104	13

He goes at some length into the equation

$$x^3-2x-5 = 0.$$

There is clearly a root a little greater than 2. Reducing by that amount,

$$x^3+6x^2+10x-1 = 0.$$

This has a root slightly less than 1. He tries 0·09. He then covers up most of his work, finally coming out with the answer

$$x = 2 \cdot 094551481542326590,$$

'correct to the 18th decimal place at three approximations.'[†]

I must confess to being somewhat sceptical as to the limits of accuracy of his methods of shortening. The general scheme is like shortened division, where we cut off figures at the beginning of the divisor, instead of adding to the end of the dividend. He states his general principle as follows:

'From these principles we make the following conclusions demonstrative of the facilities introduced by the improvement of the original process. Whatever be the dimension (n) of the proposed equation whose root is to be determined to a certain number of places, only $1/n$th part of that number (reckoning from the point at which the place of the closing addend begins to advance to the right of that part of the derivee) needs to be found by means of the process peculiar to the complete order of the equation, after which $1/n(n-1)$th may be found by the process of the $(n-1)$th order.'[‡]

† Ibid., p. 63.   ‡ Ibid., p. 64.

I am not sure exactly what this means, I am sceptical. I take

$$x^3+25x^2+5x-0\cdot961725 = 0.$$

The first figure of the root will be $0\cdot1$. Reducing by this amount

$$x^3+25\cdot3x^2+10\cdot03x-0\cdot210725 = 0.$$

Since we have advanced to the right of the last figure of the derivee, if we wish only three-figure accuracy, it would seem Horner would allow us to contract. If so, the next figure is $0\cdot02$, but if we retain the cubic this is too big.

Horner maintains that his method applies just as well when we have functions other than polynomials. He gives the example

$$x^x = 100; \qquad x\log_{10}x = 2.$$

We see from the table that a first approximation is $3\cdot6$. After this point he covers up his tracks, but it is not hard to proceed

$$\frac{d}{dx}(x\log_{10}x) = \log_{10}e(1+\log_e x)$$

$$= \log_{10}e + \log_{10}3\cdot6 + \log_{10}e\log_e\left(1+\frac{r}{3\cdot6}\right).$$

From this point on it is straightforward. He gives the astonishingly long answer†

$$x = 3\cdot597286.$$

Horner[1] is a short paper and does not call for much notice. It consists chiefly in comparing Newton's method of approximation with others not much more complicated. In Newton's method, if R be the first approximation, the first correction is $-\phi(R)/\phi'(R)$, and that is really what is usual in Horner's own method for finding roots, after things are well started. Here is something a bit more accurate. Let

$$r\phi'(R)+\phi(R) = 0,$$

$$(r+\delta r)^2\frac{\phi''(R)}{2}+(r+\delta r)\phi'(R)+\phi(R) = 0,$$

$$r+\delta r = \frac{-\phi(R)}{\phi'(R)+\frac{1}{2}\phi''(R)(r+\delta r)}.$$

We simplify the denominator: putting r for $r+\delta r$ in the denominator,

$$r+\delta r = \frac{-\phi(R)}{\phi'(R)+\frac{1}{2}r\phi''(R)}.$$

This he calls Halley's method.

In Horner[3] we get the fundamental method of Horner[2] explained in

† Horner[2], p. 68.

so simple a fashion that the author seemed a little ashamed of it. He sees that save for the later contraction he repeats the same process over and over again, each time adding a figure to his answer.

'The remark I am about to make may appear indeed, at first sight too trivial to be dwelt upon. If the formula

$$A_{n+1}y^n + B_n y^{n-1} + \ldots + K_4 y^3 + L_3 y^2 + M_2 y + N_1 \qquad (6)$$

be divided by y as many times as are represented by any one of the subscribed exponents, the several remainders will be N_1, M_2, L_3, etc., ending with that under which the selected exponent appears. Yet if we conceive y to be only a concise expression for $(x-r)$ and the formula a concise expression for

$$A_0 x^n + B_0 x^{n-1} + \ldots + K_0 x^3 + L_0 x^2 + M_0 x + N_0 = f(x) \qquad (7)$$

we find in this perfect truism the real germ for that species of transformation on which the approximate solution depends.'†

This means that to reduce the roots of $f(x) = 0$ by r we should write

$$f(x) \equiv f(y+r) \equiv A_{n+1}y^n + B_n y^{n-1} + \ldots + M_2 y + N_1.$$

The coefficients are functions of r obtained by dividing (7) by $(x-r)$, and then the quotient and so on. In a footnote to the same page we find an explanation of Horner's method of 'synthetic division' which is further explained in Horner[4] to which I shall return presently. The explanation on this page is as follows:

'Leaving the powers of x to be mentally annexed to these coefficients, and indicating in the margin the number of times r must be used to arrive at the remainder the continual division of eq. (7) by $x-r$ will assume the form

	A_0	B_0	C_0	.	.	.	K_0	L_0	M_0	N_0
	0	$A_1 r$	$B_1 r$	.	.	.	$H_1 r$	$K_1 r$	$L_1 r$	$M_1 r$
r^n	A_1	B_1	C_1	.	.	.	K_1	L_1	M_1	N_1
	0	$A_2 r$	$B_2 r$	.	.	.	$H_2 r$	$K_2 r$	$L_2 r$	$M_2 r$
r^{n-1}	A_2	B_2	C_2	.	.	.	K_2	L_2	M_2	N_2
	0	$A_3 r$	$B_3 r$	.	.	.	$H_3 r$	$K_3 r$	$L_3 r$	$M_3 r$
	A_3	B_3	C_3	.	.	.	K_3	L_3	M_3	N_3'

There is a slight confusion here because all of the A's are equal. The reason for introducing new letters is to be able to write the general formula
$$Q_e = Q_{e-1} + P_e r.$$

Horner[4] is a series of letters and other communications to the editors of *Leybourn's Repository* bearing dates between 1820 and 1827 and adds little to what I have already given. He gets a limit for positive roots by reducing all roots by 1, 2, 3, etc.—a tedious process. He gives a

† Horner[3], p. 109.

number of expedients for simplification, which do not impress me much. He becomes excited over the accusation that a certain Mr. Holdred had anticipated his work. He states emphatically: 'I am not indebted to any person that ever lived for a hint beyond what I have fairly and broadly stated in the *Phil. Trans.*'† That he had already been anticipated by Ruffini and Ch'in he clearly could not realize. In general the chief interest in Horner[4] is in connexion with synthetic division, which he explains, not very well, on pp. 44, 45. So far as I can make out this is original with him, and certainly deserving of credit even if to-day we have drifted rather far from such questions. I have already indicated his method of dividing $f(x)$ by $x-r$; let us generalize this, first illustrating multiplication. Suppose we wish to multiply

$$a_0 x^n + a_1 x^{n-1} + a_2 x^{n-2} + \dots \quad \text{by} \quad x^p + b_1 x^{p-1} + b_2 x^{p-2} + \dots$$

a_0	a_1	a_2	a_3	. . .
1	b_1	b_2	b_3	. . .
a_0	$a_1 + a_0 b_1$	$a_2 + a_1 b_1 + a_0 b_2$		. . .

Now divide

$$c_0 x^{n+p} + c_1 x^{n+p-1} + c_2 x^{n+p-2} + \dots \quad \text{by} \quad x^p + b_1 x^{p-1} + b_2 x^{p-2} + \dots$$

c_0	c_1	c_2	c_3	. .
1	$-b_1$	$-b_2$	$-b_3$	. .
$a_0 = c_0$	$a_1 = c_1 - a_0 b_2$	$a_2 = c_2 - a_1 b_1 - a_0 b_2$		. . .

§ 2. Paolo Ruffini

I now turn from Horner himself to those predecessors whom I have already mentioned. The first was Paolo Ruffini, professor in Modena, who attained celebrity by publishing one of the earliest proofs of the insolvability of the general equation whose degree is higher than four. Like Horner he started with Taylor's theorem for reducing his roots, and he also shied off from this as presumably too esoteric for pupils interested in solving equations.

'Ho dimostrato la verità di tali operazioni tanto nel NO 12 della citata Memoria come nei numeri 123 e 127 del Appendice della mia Algebra Elementare; ma siccome colà suppongo noto il calcolo differenzale, e quivi note le serie algebriche, no sarà forse inconveniente, prima di proceder inanzi, d'esporre nuovamente simile demostrazione senza supporre le indicate nozioni.'‡

The equation he would solve is

$$Ax^m + Bx^{m-1} + Cx^{m-2} + \dots + Sx^2 + Tx + U = 0. \tag{8}$$

Let $Ap + B = P_1,$

$Ap^2 + Bp + C = P_2,$

. . . . . .

$Ap^n + Bp^{n-1} + Cp^{n-2} + \dots + Lp^2 + Mp + N = P_n.$

† Horner[4], p. 39.   ‡ Ruffini (q.v.), p. 12.

These P_i's are what we get by dividing (8) by $x-p$. He then introduces a whole new flock of letters:

$$\alpha_2 = pA+P_1 \qquad \alpha_3 = p\alpha_2+P_2 \qquad \alpha_4 = p\alpha_3+P_3$$
$$\beta_3 = pA+\alpha_2 \qquad \beta_4 = p\beta_3+\alpha_3 \qquad \beta_5 = p\beta_4+\alpha_4$$
$$\gamma_4 = pA+\alpha_3 \quad . \quad . \quad . \quad . \quad . \quad . \quad . \quad .$$

These are Horner's numbers in a different order; it fatigues the memory to try to hold them. I note in passing

$$\frac{dP_n}{dp} = \alpha_n; \qquad \frac{d^2P_n}{dp^2} = \beta_n.$$

He finally sets out the array

$$
\begin{array}{c|cccccc}
p & A & B & C & D & E & F \\
& A & P_1 & P_2 & P_3 & P_4 \\
& & A & \alpha_2 & \alpha_3 & \alpha_4 \\
& & & A & \beta_3 & \beta_4 \\
\end{array}
$$

which is Horner's essential arrangement.

Ruffini goes to great lengths to find methods of contraction, but I see no reason to follow him there. I repeat that it seems to me very unlikely that Horner had ever heard of this paper.

§ 3. Ch'in Chiu-Shao

A far earlier predecessor of Horner's, whom it is even less likely that he ever heard of, was Ch'in Chiu-Shao, a thirteenth-century Chinese sage who sometime around A.D. 1247 composed the *Nine Sections of Mathematics*. Here, as far as I can understand it from Mikami† is his general scheme for the solution of numerical equations.

A first approximation to a root is guessed. The first coefficient is multiplied by the approximate root and added to the second coefficient, making the first partial sum of the first set. The first coefficient is multiplied by the approximate root and added to the first partial sum, making the second partial sum of the first set, and this system is kept up till a first set of n partial sums is made. Then we start again. The first partial sum is multiplied by the approximate root and the product is added to the third coefficient, making the first partial sum of the second set. We keep on in this way, multiplying the partial sums of the first set by the approximate root, and adding to the corresponding partial sums of the second set. This is done $n-1$ times, making the second set of partial sums, and so on. This is evidently Horner's

† q.v., pp. 74 ff.

reduction of all roots by the amount of the approximation, using his method of synthetic division in reversed order. Here is a simplification of Ch'in's process as given by Mikami.† The table should be read up from the bottom. The equation to be solved is

$$-x^4 + 763200x^2 - 40642560000 = 0.$$

If we solve this as a quadratic in x^2 we find four real approximate roots $x = \pm 76$, $x = \pm 265$. But Ch'in, for some reason I do not grasp, takes as his first approximation 800.

		800	Root
$(1°) \times 800$	800×98560000	-40642560000 78848000000	
		38205440000	Absolute
$(2°) \times 800$	800×123200 $800 \times (-1156800)$	98560000 -925440000	
		-826880000	1st degree
$(3°) \times 800$	$800 \times (-800)$	763200 -640000	
		123200	
	$800 \times (-1600)$	-1280000	
		-1156800	
	$800 \times (-2400)$	-1920000	
		-3076800	2nd degree
$(4°) \times 800$	$800 \times (-1)$ $800 \times (-1)$	-800 -800	
		-1600	
	$800 \times (-1)$	-800	
		-2400	
	$800 \times (-1)$	-800	
		-3200	3rd degree
		-1	4th degree

I repeat that we have here essentially Horner's own method.

I think that my partiality for Horner is partly based on a prejudice in favour of a man of limited training who, by courage and determination, did something really worth while, even if others had done it before him. But no subsequent writer has developed anything better, his fundamental idea and his synthetic division are the last word in simplicity; greater mathematicians have done less admirable things.

† q.v., pp. 74–7.

CHAPTER XVI

BERNHARD BOLZANO

§ 1. Roots of a real function

THE distinguished Czech writer whose name stands above bears, from the point of view of this book, a certain resemblance to Blaise Pascal. Both were deeply interested in religion and philosophy, both were involved in controversy and suffered for the faith that was in them. Both did such brilliant mathematical work that they might well be classed as professionals. But I have included Pascal with the amateurs because he was more famous as a philosopher and a writer of beautiful French prose than as a mathematician, and I take up Bolzano because it seems to me interesting that a man who was a remarkable pulpit orator, only removed from his chair for his political opinions, should have thought so far into the deepest problems of a science which he never taught in a professional capacity. An interesting study of Bolzano the mathematician will be found in Stolz (q.v.).

Let us begin by noticing Bolzano[1], which was completed in 1817, and bears the title *Rein analytischer Beweis des Lehrsatzes, dass zwischen je zwey Werthen, die ein entgegengesetztes Resultat gewähren, wenigstens eine reelle Wurzel liege.* This tells us that if the real function $f(x)$ be continuous through the interval from a to b, and if $f(a)$ and $f(b)$ have opposite algebraic signs, then at least one root lies in the interval. This fact is so intuitively evident if one draws a picture, that it is hard to persuade any but a professional mathematician that there is any interest in proving it.

But the words *rein analytischer Beweis* show that Bolzano looked on this as a purely analytical fact, which should be proved by steps based merely on the premises of analysis. The date is important because the work contains the earliest statement of two most important mathematical principles, and because it was written by a man still actively engaged in non-mathematical work, who must have had but a limited amount of time to give to mathematical research.

He begins with the statement that two theorems have occupied a very central place in the study of algebra. The first is the theorem stated above, the second that which states that any polynomial in one variable with real coefficients is divisible into factors which are either linear or quadratic in the variable. This second theorem had been adequately handled by Gauss; as for the first, there were many existing proofs, but either they were based on geometry, that is to say on pure intuition, or on insufficiently exact ideas of continuity. We should have a really rigorous proof, based on accepted principles germane to the theorem involved.

Bolzano[1] is occupied with real functions of real variables, especially with continuous functions. But when is a function continuous ? Here is his definition:

'Eine Function $f(x)$ für alle Werthe von x die inner- oder ausserhalb gewisser Grenzen liegen, nach dem Gesetze der Stetigkeit sich ändere, nur so viel, dass, wenn x irgend ein solcher Werth ist, der Unterschied $f(x+w)-f(x)$ kleiner als jede gegebene Grösse gemacht worden könne, wenn man w so klein, wie man nur immer will, annehmen kann.'†

This definition of continuity, which would be perfectly acceptable if he indicated that *klein* means small in absolute value, and which is absolutely fundamental in all modern discussions of real variables, is often attributed to Cauchy, but is really Bolzano's own great contribution to analysis.‡ If Bolzano had done nothing else in mathematics, this alone would secure for him a place in the history of the subject.

We first get down to business in Bolzano[1] with the study of infinite sequences. On p. 21 of Bolzano[1] we find a statement which I translate as follows.

Theorem 1] *When a series of values $F_1(x)$, $F_2(x)$, $F_n(x)$,..., $F_{n+r}(x)$ is such that the difference between the nth term $F_n(x)$ and every subsequent term $F_{n+r}(x)$, be this never so far removed, remains less than any specified number when n is taken sufficiently large, then there is always a definite value X, and only one, to which the members of the series approach as close as we will, when the series is prolonged indefinitely.*

This is of capital importance in analysis. X is not a constant, but a function of x, and we have here the first adequate definition of a convergent series.§ Moreover, he does not give a real proof at this point; I will return to this later on. Bolzano points out that such convergent series exist, as the series of sums in the series

$$1+r+r^2+... \qquad |r| < 1.$$

He also shows that $F_n(x)$ itself approaches this value as

$$|X-F_n(x)| \leqslant |X-F_{n+r}(x)|+|F_{n+r}(x)-F_n(x)|,$$

and similarly we see that there can be only one such value. What is really lacking is an existence theorem for X unless we take the word *Lehrsatz* to mean assumption. He has no real definition of the continuum, or of irrational number. The series $1+0.4+0.01+...$ converges

---

† Bolzano, pp. 7, 8.

‡ Cf. Jourdain in Bolzano[1], p. 39, and Pringsheim in the *Encyclopädie der math. Wissenschaften*, vol. ii, 1. 1.

§ Cf. Jourdain in Bolzano[1], p. 41, and Knopp (q.v.), p. 85.

to $\sqrt{2}$, but what sort of a thing is this to which we say it converges? A clear definition of an irrational number by a Cantor series or a Dedekind cut is still some distance in the future. What is lacking here, as elsewhere in his published work, is a proof of the Bolzano–Weierstrass theorem: *Every bounded sequence possesses at least one limiting point.*[†] I shall return to this again.

From this point on Bolzano[1] runs smoothly enough. If a series be such that by taking n large enough the sum of all the terms after n of them can be made as small as we please in absolute value, the series of sums is convergent, as is the series itself.

Theorem 2] *If not all values of a variable have a certain property M, but all which are less than a given value u have this property, then there is a certain value U such that it is the greatest number of which it can be said that all lesser values of x have this property.*

Here is his proof. Let all values of x for which this holds be called values of the first class, and all others of the second class. We can find D, when U is of the first class, such that $U+D$ is of the second. Consider the values

$$U+\frac{D}{2^0}, \quad U+\frac{D}{2^1}, \quad U+\frac{D}{2^2}, \quad ..., \quad U+\frac{D}{2^l}.$$

If no such value is of the first class, no matter how large l may be, as this number can be made as close to U as we please, then U itself is the number we seek. If not, let $U+(D/2^n)$ be the first of these which is of the first class, while $U+(D/2^{n-1})$ is of the second. Using $D/2^n$ in the way that we previously used D, and continuing the same process again and again, we have

$$U+\frac{D}{2^n}+\frac{D}{2^{n+m}}+...+\frac{D}{2^{n+m+...+r}}.$$

If this series terminate it ends in the sort of number we want; if not, it tends to a limit under Theorem 1.[‡]

Theorem 3] *If two functions of x, $f(x)$, $\phi(x)$ either for all values of x or for all between α and β alter according to the law of continuity, and if further $f(\alpha) < \phi(\alpha)$; $f(\beta) > \phi(\beta)$, then there is always between α and β such a value that*

$$f(x)-\phi(x) = 0.$$

We suppose both α and β are positive, and $\alpha < \beta$. We shall say that w has the property M, if $f(\alpha+w) < \phi(\alpha+w)$. When $w = 0$ it has this property, but not when it $= \beta-\alpha$. Then by Theorem 2 there is a largest value of w such that for $x < \alpha+w, f(x) < \phi(x)$. By the continuity

[†] Cf. Knopp (q.v.), p. 89.

[‡] Jourdain says, Bolzano[1], p. 43, 'Die folgende Schlussweise . . . ist im Wesen sehr alt', but gives no reference.

assumption, for this value $f(x) = \phi(x)$. The fundamental theorem he seeks to prove is a special case of this when $\phi(x) \equiv 0$.

§ 2. Study of real functions

Bolzano[2] is a careful study of real functions of a real variable, the domain being the continuum. He builds from the ground up. It is a more elaborate and sophisticated work than Bolzano[1], showing a much greater knowledge of the literature; he doubtless had read Cauchy. He begins with a study of differences which is not particularly interesting. Matters improve when he passes to continuous functions, the definition being essentially that of Bolzano[1]; but this time he speaks of the absolute values of his differences. He points out that a function may be continuous at a point for a positive value of the increment only. In general he only uses the word *stetig* when there is continuity both ways. He occasionally goes astray, writing

$$\frac{1}{1+x} = 1-x+x^2-x^3+\dots.$$

Then 'by the law of continuity'

$$\tfrac{1}{2} = 1-1+1-1+\dots.$$

A slip occurs on p. 27, where he assumes that if $F(x)$ be continuous in a certain interval, and take the value M an infinite number of times, without being identically equal thereto, then to each point where $F(x) = M$ there is a nearest point where the same is true. Rychlik[†] takes the function $F(x) = (1/x)\sin x$ $(-1 \leqslant x \leqslant 1)$, $F(0) = 0$. There is no nearest point to the origin where $F(0) = 0$. Something very interesting comes on pp. 28, 29.

Theorem 4] *If a function $F(x)$ defined for an infinite number of values between a and b take an infinite number of values, and if at least one of these be greater than any assigned number, this function is not continuous for all values of the interval.*

We assume $x_1, x_2, x_3,\dots$ such that

$$F(x_1) > 1, \quad F(x_2) > 2, \quad \dots, \quad F(x_n) > n, \quad \dots.$$

Then there must be a point of accumulation c in the interval such that in the shrinking interval $(c+w)$, $F(x)$ takes a value greater than any assigned quantity.

But why must there be such a point of accumulation? Bolzano writes in a footnote in his own handwriting that this was shown in the 'Lehre von der Messbarkeit der Zahlen'. But where is this article, and when was it written? Rychlik says in a footnote to Bolzano[2][‡] that neither he nor Jašek has been able to find any trace of such a work.

† Bolzano[2], p. 5.   ‡ Bolzano[2], Notes, p. 5.

Now this is exactly the Bolzano–Weierstrass theorem that I mentioned on p. 197 which is used without proof in Bolzano[1]. It would be pleasant to assume that Bolzano himself worked out a rigorous demonstration, but we can hardly take this as certain.

We pass from here, by similar reasoning, based on this theorem to

Theorem 5] *If $F(x)$ be continuous in a closed interval a to b it will take at least once every value between $F(a)$ and $F(b)$.*

Theorem 6] *If $F(x)$ be continuous in a closed interval a to b it will actually take a maximum and a minimum value.*

He passes next to functions of several variables, and is at pains to define exactly what he means by continuous. He gives

Theorem 7] *If for all values of x, y, z,... in the closed interval $x\pm h$, $y\pm k$, $z\pm l$ the function $F(x, y, z,...)$ is continuous in each of the variables separately, it is continuous in all together.*

$$F(x+\Delta x, y+\Delta y, z+\Delta z,...) - F(x, y, z,...) = F(x+\Delta x, y+\Delta y, z+\Delta z,...) -$$
$$- F(x, y+\Delta y, z+\Delta z,...) + F(x, y+\Delta y, z+\Delta z,...) - F(x, y, z+\Delta z,...) +$$
$$+ F(x, y, z+\Delta z,...) - F(x, y, z,...) + ...$$

Each of these separate differences approaches 0.

Rychlik points out that this is true if we assume that this limiting value is independent of the manner of approach, otherwise it may not be true.

Let
$$F(x, y) = \frac{xy}{x^2+y^2},$$
$$F(0, y) \equiv 0, \qquad F(x, 0) \equiv 0.$$

If
$$y = mx, \qquad F(x, y) = \frac{m}{1+m^2}.$$

He then returns to functions of one variable and shows that even though in a closed interval a to b, $F(x)$ takes all intervening values, it is not necessarily continuous. For instance we might take the interval 0 to 1, making $F(x) = \frac{1}{2}x$, x rational, $F(x) = x$, x irrational, $F(0) = 0$, $F(1) = \frac{1}{2}$.

He studies monotonely increasing or decreasing functions and functions with an infinite number of maxima or minima in a closed interval. Rychlik finds several flaws in the proofs, but I will not pause to point them out.

The second section of Bolzano[2] is given to derivatives. He is careful to point out that there may be a difference in value between the derivatives in the positive and in the negative direction. If a function have a derivative, either in the positive or the negative direction, for a value $x = x_0$, it is continuous in this direction; the converse is not necessarily true.

Theorem 8] *If a function $F(x)$ is continuous at x_0 but has no derivative there, then either $\Delta F/\Delta x$ increases indefinitely in absolute value, or there is such a number M that $|M-\Delta F/\Delta x|$ can be made as small as we please.*

We see, in fact, that if $\Delta F/\Delta x$ does not increase indefinitely it must remain less than some smallest number M. He gives some attention to functions nowhere differentiable, and has a function with an infinite number of maxima and minima which is too complicated to give here. He pays some attention to the Law of the Mean, and Taylor's theorem with remainder. The proofs are inferior to others now available, and there are occasional lapses in the reasoning.

Bolzano[2] is, as I said before, a more sophisticated work than Bolzano[1] and is based on a wider mathematical knowledge. The number of small faults is large; it would not do to-day as an introduction to the subject. But it is a remarkably penetrating study of a very difficult subject, a noteworthy work for any amateur.

Bolzano[3], like Bolzano[2], is carefully built up from first principles. It deals largely with divisibility and primality of positive integers, a far safer subject for an amateur than the general theory of real functions. He proves various standard theorems, such as Euclid's method of finding the highest common factor of two integers, the unicity of division into prime factors, one of Fermat's famous theorems, unfortunately not his last, and Wilson's theorem. A good deal of attention is given to the theorem that every integer is the sum of four or fewer perfect squares. There are few errors as compared with Bolzano[2]. This is quite natural, given the difference in difficulty of the two subjects. There does not seem to be anything very original in the whole work, it is not a subject where original results are easily found.

§ 3. *Die Paradoxien des Unendlichen*

Bolzano[4] represents a serious attempt to wrestle with the difficulties inherent in the introduction of infinitely large quantities into mathematics. How many great men, even great mathematicians, have attempted this, and with what pitiful results! Prihonsky says in the introduction, p. iv: 'Wahrhaftig hätte Bolzano nichts anderes geschrieben, und uns hinterlassen, als diese Abhandlung allein, er müsste, wie wir fest glauben, schon um ihretwillen den ausgezeichnetsten Geistern unseres Jahrhunderts beigezählt werden.' Georg Cantor (q.v.) expresses an almost equally high appreciation of Bolzano's penetration, though he points out places where the Czech theologian falls short of the mark.

Bolzano's thesis is that the infinite is something perfectly definite, with definite properties, not merely a pure negation. He proposes to deal with some of the surprising results, even apparent paradoxes, which arise from this conception. His first approach is through the ordinal numbers. Suppose we have an assemblage where we can say of any two things either that they are equal, or that one is greater than the other, and that to each which is not the greatest of all, if there be such, there is one next greater. Suppose, further, there is one least. Then to each object will correspond a positive integer. If there is no greatest we say that the assemblage is infinite. He says that he has no quarrel with writers such as Cauchy, who define the infinite as a variable and an assemblage as infinite if it can be increased indefinitely, though he rightly objects to those who call infinity the limit of an unlimited assemblage.

Bolzano has more difficulty with infinite cardinals. He remarks that there is hardly a mathematician who will deny that the number of points on a line between two given points is infinite, but it cannot be said that in this continuum there is one point which is next to another. He points out next that a finite object can be composed of an infinite assemblage of parts; an example is the sophists' paradox of Achilles and the tortoise, or an endless decimal. And now comes his first and greatest paradox. All assemblages are not equal, but the ancient axiom that a whole is greater than a part of itself must be modified because an infinite assemblage is equivalent to a part of itself in the sense that they can be put into one-to-one correspondence. It is easy to show examples. The points of the X-axis whose abscissae are 0 or positive can be put into one-to-one correspondence with those whose abscissae are greater than or equal to any chosen positive number. Or again, if we take the two ordered sequences

$$1, \quad 2, \quad 3, \quad 4, \quad ...,$$
$$1^2, \quad 2^2, \quad 3^2, \quad 4^2, \quad ...$$

there is a one-to-one correspondence between them, but the second sequence contains only a minor portion of the first.

Here Bolzano stands on the brink of taking a great forward step, that of describing the *Mächtigkeit* or power of an infinite assemblage, and showing that not all infinite assemblages are equivalent in the sense that they can be put in one-to-one correspondence with one another. He was not quite able to take it, we must wait for Georg Cantor, but he was very near. He writes:

'Übergehen wir nun zur Betrachtung einer höchst merkwürdigen Eigenheit, die in dem Verhältnisse zweier Mengen, wenn beide unendlich sind, vorkommen kann, ja eigentlich immer vorkommt, die man aber bisher zum Nachtheil für die Erkentniss mancher wichtigen Wahrheiten der Metaphysik, sowohl als Physik und Mathematik übersehen hat, und die man wohl jetzt, indem ich sie aussprechen werde, in einem solchen Grade Paradox finden wird, dass es sehr nöthig sein dürfte, bei ihrer Betrachtung uns etwas länger zu verweilen.'[†]

† Bolzano[4], p. 28.

This seems to be nothing less than a claim that no one had previously noticed that the members of an infinite assemblage can be put in one-to-one correspondence with those of a part of the same assemblage. It is indeed hard to believe that no previous writer had ever noticed this fairly evident fact, but I can find no mention of it, one way or the other, in the literature. It would be pleasant to think that it was Bolzano's own original contribution to fundamental mathematics. I will mention in passing that it can be taken as a definition for the infinitude of an assemblage.

At this point Bolzano wanders off into some calculations which, as Georg Cantor† rightly says, do not make any sense at all. He writes:

$$1^0+2^0+3^0+...+n^0 = N^0.$$

What the sign of equality means here I do not know. Then we have

$$(n+1)^0+(n+2)^0+(n+3)^0+... = N^n$$
$$1^0+2^0+3^0+... \qquad\qquad = N^n-N^0$$
$$1+2+3+...+n+(n+1)... \quad = S^0,$$

but adds we may not write

$$S^0 = \frac{N^0(N^0+1)}{2}$$

merely because this is true of finite numbers.

A number is said to be infinitely small if no matter how many times we take it, we cannot come up to a finite number. He does not show that such things exist, or in what sense we here use the word number. He points out the danger of writing $1/N^0$ for such a number, and the still greater danger at the other end of writing $1/0$. The series

$$x = a-a+a-a+...$$

must not be written

$$x = a-(a-a)-(a-a)...,$$

as this leads to $x = a/2$. He points out that in an infinite series we may not change the order of terms, or introduce parentheses as we do with finite sums. If we write

$$\frac{a}{1-1} = a+a+a+...+\frac{a}{1-1},$$

$$a+a+a+... = 0.$$

I cannot help expressing surprise that the man who first introduced the correct definition of a convergent series should not have grasped the idea of defining a series as a limit.

† q.v., p. 561.

Bolzano points out the great danger of following Leibniz in neglecting small numbers in comparison with large ones. If M and N are two large numbers whose ratio is rational we cannot say of any number μ however small that

$$\frac{M+\mu}{N} = \frac{M}{N}.$$

Bolzano now passes to the safer ground of the infinitesimal calculus. If y be a function of x which has a derivative, we are safe in writing

$$\lim \frac{\Delta y}{\Delta x} = \frac{dy}{dx}.$$

When it comes to second differences he proceeds like this:

$$y+\Delta y = y+\frac{dy}{dx}\Delta x+\frac{1}{2}\frac{d^2y}{dx^2}\Delta x^2+...;$$

$$(y+2\Delta y+\Delta^2 y) = y+\Delta y+\frac{d(y+\Delta y)}{dx}\Delta x+\frac{1}{2}\frac{d^2(y+\Delta y)}{dx^2}\Delta x^2+...;$$

$$\Delta y = \frac{dy}{dx}\Delta x+\frac{1}{2}\frac{d^2y}{dx^2}\Delta x^2+...;$$

$$\frac{d\Delta y}{dx} = \frac{d^2y}{dx^2}\Delta x+...;$$

$$\Delta^2 y = \frac{d^2y}{dx^2}\Delta x^2+...;$$

$$\lim \frac{\Delta^2 y}{\Delta x^2} = \frac{d^2y}{dx^2}.$$

This, of course, is perfectly sound.

The continuum of time and space is defined as everywhere dense. Lengths are measured in comparison with other lengths, areas with areas, and so on. He pays a good deal of attention to the erroneous idea that the distance between two points is measured by the infinite number of points between them. A curious result of this would be that the ratio between the number of points on a side and on the diagonal of a square would be irrational, or more simply, as the points on a line segment an inch long can be put into one-to-one correspondence with those on a line 2 inches long, the two lengths are equal. There is also much danger in speaking of infinitely short distances. Here is something curious; I name it after the man Bolzano said discovered it:

Galileo's Paradox] *The circumference of a circle has the same area as its centre.*

The modern mathematician would agree, as each has the area 0, but let us see what is meant. Consider a sphere of radius 1, embedded in a circular cylinder of that radius and height 2. Consider a ring in the plane $z = z_1$ between two circles with a common centre $(0, 0, z_1)$ and radii 1 and $\sqrt{(1-z_1^2)}$: $\pi z_1^2 = \pi \cdot 1 - \pi(1-z_1^2).$

The limit of the left-hand side as $z_1 \to 0$ is the centre, and of the right is the equator.

Bolzano next goes on to show that certain things, formerly considered paradoxical, are not really so. It was formerly believed that an infinitely extended object could not be contained in a continuum of finite extent, or that an infinitely extended object could not have a finite content, and that a spiral with an infinite number of turns must be infinitely long. None of these assumptions is correct.

Consider first the curve

$$y = \sin\frac{1}{x},$$

or so much of it as is contained between the Y-axis, the line $x = 2/\pi$, and the lines $y = \pm 1$. The length from the origin to the line $x = 2/\pi$ is infinite. Next consider the curve

$$x^3 y^2 = 1.$$

This will be extended indefinitely along the X-axis. The area between the curve and that axis from the point $(1, 1)$ out is

$$\int_1^\infty y \, dx = \int_1^\infty x^{-\frac{3}{2}} \, dx = 2.$$

Bolzano's example is not so simple as this. Before leaving this curve let us look at the other asymptote. The area here is

$$\int_1^\infty x \, dy = \int_1^\infty y^{-\frac{2}{3}} \, dy = \infty.$$

But if we spin around this axis the volume generated is

$$\pi \int_1^\infty x^2 \, dy = \pi \int_1^\infty y^{-\frac{4}{3}} \, dy = 3\pi.$$

Query, what has become of Pappus' theorem that the volume generated by spinning an area about a line in its plane which does not cross it, is equal to the product of the area multiplied by the length of the path traced by the centre of gravity? As a mistake of the statement about the spiral with an infinite number of turns, we have but to consider the logarithmic spiral which turns an infinite number of times as we go in towards the centre, but has a finite limit of length.

Here is one more example of the fundamental theorem that an infinite assemblage is equivalent to a part of itself. If two figures can be carried into one another by a rigid motion of the plane, it would seem safe to say that they included the same infinite number of points. It is equally safe to say that there are as many positive numbers as negative ones. Similarly it would seem safe to say that the number of numbers < 2 is equivalent to the number of numbers > 2, for the last statement is carried into this by sliding the origin 2 places. But the number of numbers < 2 is greater than the number of negative numbers, and so greater than the number of positive numbers and so greater than the number > 2. What we really need here is Bernstein's theorem, still far in the future, that two assemblages are equivalent when each is equivalent to a part of the other.

The remainder of this interesting essay is more or less metaphysical and so is not our present concern. The whole is the work of a man who reflected deeply on the foundations of mathematics, and came very near to developing the modern theory of infinite assemblages. But what a splendid accomplishment for one who was not a professional mathematician to be the first to define properly a continuous function, the first to define properly a convergent sequence, and the first to point out that an infinite assemblage is equivalent to a part of itself.

INDEX OF AUTHORS QUOTED

Alberti, Leon Battista, 1404–1472, *The Painting of Leon Battista Alberti*, London, 1726, 31–2.

Amadeo, Frederico, 1859– , 'Albrecht Dürer precursore di Monge', *Atti della R. Accademia di Napoli* (2), 13, 1908, 69–71.

Archimedes[1], 287–212 B.C., *Opera omnia*, Heiberg ed., Leipzig, 1881, 11, 22.

Archimedes[2], *The Works of Archimedes*, translated by T. L. Heath, Cambridge, 1897, 21, 24, 25, 71.

Archimedes[3], *The Method of Archimedes*, translated by T. L. Heath, Cambridge, 1912, 43.

Arnauld[1], Antoine, 1612–1698, *Logique de Port-Royal*, avec introduction et des notes par Charles Jourdain, Paris, 1854, 103–8.

Arnauld[2], *Nouveaux Élémens de géométrie*, Paris, 1667, 108–18.

Ball, Walter William Rouse, 1850–1925, *A Short Account of the History of Mathematics*, London, 1901, 73, 170.

Barnwell, Robert Gibbs, *Life and Times of John De Witt*, New York, 1856, 127–31.

Benecke, Adolph, *Über die geometrische Hypothesis in Platon's Meno*, Elbing, 1867.

Bernoulli, John, 1677–1748, *Die Differentialrechnung*, Ostwald's *Klassiker der exakten Wissenschaften*, Leipzig, 1924.

Birch, Thomas, 1705–1788, *History of the Royal Society of London*, London, 1751, 141.

Bolzano[1], Bernhard, 1781–1848, *Rein analytischer Beweis*, etc., Ostwald's *Klassiker der exakten Wissenschaften*, Leipzig, 1908, 193–7.

Bolzano[2], *Schriften*, vol. i, *Funktionentheorie*, published by the Royal Czech Scientific Society, Prague, 1930, 198–200.

Bolzano[3], *Schriften*, vol. ii, *Zahlentheorie*, ibid., 200.

Bolzano[4], *Die Paradoxien des Unendlichen*, Berlin, 1889, 200–5.

Bopp[1], Karl, 'Antoine Arnauld als Mathematiker', *Abhandlungen zur Geschichte der math. Wissenschaften*, vol. xiv, Leipzig, 1902, 103, 108, 109, 117, 142.

Bopp[2], 'Die Kegelschnitte des Gregorius a St. Vincento', *Abhandlungen zur Geschichte der math. Wissenschaften*, vol. xx, Leipzig, 1907, 142.

Bortolotti, Ettore, 1866– , 'La scoperta delle frazioni continue', *Bollettino della Mathesis*, Anno xi, 1919, 138.

von Braunmühl, A., *Vorlesungen über Geschichte der Trigonometrie*, Leipzig, 1903, 81–2.

Brianchon, Charles Julien, 1783–1864, 'Les Courbes de deuxième degré, *Journal de l'École Polytechnique*, vol. vi, 1806, 90.

Brouncker[1], William, Viscount, 1620–1684, 'The Squaring of the Hyperbola', *Phil. Trans. abridged*, vol. i, 1809, 137–9.

Brouncker[2], 'Demonstration of the Synchronism of the Vibrations in a Cycloid', *Phil. Trans. abridged*, vol. ii, 1909, 141.

Leonardo[4], 'I Manoscritti ed i Disegni di Leonardo da Vinci', *Il Codice Arundel*, Rome, 1923, 43, 58–9.

L'Hospital[1], Guillaume François Antoine, Marquis de Sainte-Mesme, *Analyse des infiniment petits*, Paris, 1696, 154–63.

L'Hospital[2], *Traité analytique des sections coniques*, Paris, 1707, 163–70.

Libri, Guillaume, 1803–1869, *Histoire des sciences mathématiques en Italie*, Paris, 1838, 43–4, 50.

Lilly, William, 1602–1681, *A History of his Life and Time*, London, 1715.

Mancini, Giralomo, 1842–1932, 'De corporibus regularibus di Pietro Franceschi detto Della Francesca', *Atti della R. Accademia dei Lincei (Morali)*, (5), vol. xiv, 1915, 42.

Marcolongo[1], Roberto, 1862–   , *Studi Vinciani*, Naples, 1937, 43–59.

Marcolongo[2], *Leonardo da Vinci, Artista Scienziato*, Rome, 1939, 43, 45.

Marie, F. Maximilien, 1819–1891, *Histoire des sciences mathématiques*, Paris, 1883–1888, 89.

Mikami, Yoshio, 'The Development of Mathematics in China and Japan', *Abhandlungen zur Geschichte der Mathematik*, vol. xxx, Leipzig, 1913, 86, 193–4.

Montucla, Jean Étienne, 1725–1789, *Histoire des mathématiques*, Paris, 1800, 143.

Napier[1], John, Baron of Merchiston, 1550–1617, *The Construction of the Wonderful Canon of Logarithms*, by John Napier, translated by R. Macdonald, Edinburgh, 1889, 73–6, 81–2.

Napier[2], *The Napier Tercentenary Memorial Volume*, London, 1915, 71–5, 78, 86–8.

Napier[3], *Logarithmorum Canonis Descriptio*, London, 1620, 80.

Napier[4], *Rabdologiae sive Numerationis per virgulas libri duo*, Edinburgh, 1617, 82, 84.

Napier[5], *De Arte Logistica*, Johannis Napier, Edinburgh, 1839, 85–8.

Narducci, Enrico, 1832–1893, *Intorno ad una traduzione italiana fatta nel secolo XIV del trattato di ottica*, 53.

Omar Khayyám, 1100, *L'Algèbre de Omar Alkhayyami*, par Franz Woepcke, Paris, 1851. Ch. II.

Oznam, *Dictionnaire de Mathématiques*, Paris, 1691.

Paccioli, 1440–1519, 'De divina Proportione', translated by Constantin Winterberg, *Quellenschnitte für Kunstgeschichte*, Neue Folge, Vienna, 1889, 41, 44.

Pappus, d'Alexandrie, 300, *La Collection mathématique*, translated by Ver Ecke, Paris, 1933, 22.

Paradies, Ignace Gaston, 1636–1673, *La Statique*, Paris, 1673, 141.

Pascal, Blaise, 1623–1662, *Œuvres complètes*, Édition Brunschwicg, Paris, 1923, Ch. VII.

Perron, Oskar, 1880–   , *Die Lehre der Kettenbrüche*, Leipzig, 1929, 138.

Pittarelli, 'Intorno al libro De prospettiva pingendi', *Atti del Congresso Internazionale di Scienze storice*, Rome, 1904, 33.

Plato[1], 430–349 B.C., *The Dialogues of Plato*, translated by Benjamin Jowett, 3rd ed., Oxford, 1892. Ch. I.

Plato[2], *The Republic of Plato with notes and commentaries*, by James Adam, Cambridge, 1906, 7.

Plutarch, 46–125, *Plutarch's Lives, the translation called Dryden's*, revised by Arthur Hugh Clough, London, 1859, 12.

Proclus, 410–485, *The Philosophical and Mathematical Papers of Proclus*, translated by T. Taylor, London, 1792, 1, 4, 12, 16, 124.

Ramée, Pierre de La, 1515–1572, *Via Regia ad Geometriam*, translated by William Bedwell, London, 1620, 118.

Rebel, Otto Julius, 1910– , *Der Briefwechsel zwischen Johann Bernoulli und dem Marquis L'Hospital*, Bottrop, 1934, 147–8, 161, 163.

Recorde, Robert, 1510–1558, *The Pathway to Knowledge*, 2nd ed., London, 1574, 118.

Reiff, Richard, 1855– , *Geschichte der unendlichen Reihen*, Tübingen, 1889, 137.

Routh, Edward John, 1831–1907, *Elementary Rigid Dynamics*, London, 1913, 184.

Ruffini, Paolo, 1765–1822, 'Un nuovo metodo generale di estrare le radici numerali', *Società Italiana delle Scienze*, vol. xvi, 1801, 182, 193.

Schafheitlein, Paul, 1861– , *Johann Bernoulli, die Differentialrechnung*, Leipzig, 1924, 161–3.

van Schooten[1], Franz, 1615–1660, *Geometria a Renato Descartes*, Amsterdam, 1683, 119–27, 166.

van Schooten[2], *Exercitationum Mathematicarum*, Leyden, 1657, 120, 124–5.

Schwenter, Daniel, 1585–1636, *Geometriae practicae novae et auctae Tractatus*, Nuremberg, 1626, 138.

Scott, J. F., *The Mathematical Work of John Wallis*, Oxford, 1938, 136.

Smith, David Eugene, 1860–1944, *History of Mathematics*, vol. i, Boston, 1923, 73–4, 85, 147.

Stifelius, Michaelis, 1487–1567, *Arithmetica Integra*, Nuremberg, 1543, 72, 86, 93.

Stolz, Otto, 1842–1903, 'Bolzano's Bedeutung in der Geschichte der Infinitesimalrechnung', *Math. Annalen*, vol. xviii, 1888, 195.

Suter, Heinrich, 1848–1922, 'Die Kreisquadratur des Ibn al-Haitan', *Zeitschrift für Mathematik und Physik*, vol. xliv, 1899, Historical Part, 45.

Tropfke, Johannes, 1866–1939, *Geschichte der Elementarmathematik*, 2nd ed., Berlin, 1921; 3rd ed., 1930, 20, 81, 86, 114–5, 135, 139.

Valla, Giorgio,   –1500, *De expetendis et fugiendibus rebus*, 1501.

Vasari, Giorgio, 1512–1574, *Lives of Seventy of the Most Eminent Painters and Sculptors*, New York, 1909, 30.

Vieta, Franciscus, 1540–1603, *Opera Mathematica in unum volumen conjecta*, Lyons, 1646, 80.

Vitruvius, 1st century B.C., *The Ten Books of Architecture*, translated by M. H. Morgan, Cambridge, Mass., 1914, 30.

Wallis, John, 1616–1703, *Opera mathematica*, Oxford, 1695, 136–7, 139.

Waters, William George, 1893– , *Piero della Francesca*, London, 1908, 41.

Wiedemann, Eilhard, 1852–1928, 'Ibn Al Haitan's Schrift über die sphärischen Hohlspiegel', *Bibliotheca Mathematica* (3), vol. x, 1910, 53.

Woepcke, Franz, 1826–1864, 'Histoire des mathématiques chez les Orientaux', *Journal Asiatique* (5), vol. v, 1855, 55, 57.

Zeuthen, Hieronymus, 1839–1910, 'Geschichte der Mathematik im XVI. und XVII. Jahrhundert', *Abhandlungen zur Geschichte der mathematischen Wissenschaften*, vol. xvii, Leipzig, 1903, 144, 146.